CLASS 08 LOCOMOTIVES IN SCOTLAND

COLIN J. HOWAT

AMBERLEY

First published 2023

Amberley Publishing
The Hill, Stroud
Gloucestershire, GL5 4EP

www.amberley-books.com

ISBN 978 1 3981 0560 7 (print)
ISBN 978 1 3981 0561 4 (ebook)

British Library Cataloguing in Publication Data.
A catalogue record for this book is available from the British Library.

Typesetting by SJmagic DESIGN SERVICES, India.
Printed in the UK.

Introduction

Once a common sight in most Scottish towns and cities, Class 08 shunting locos could be found virtually anywhere from Dumfries to Inverness. With the introduction of more unit stock and the rundown of freight operations, the Class 08 is sadly a rare beast nowadays. Only a few locations in Scotland continue to use them, and this is mainly at depots such as Craigentinny and Inverness. Standard diesel locos such as the Class 66 now perform the job of marshalling trains on arrival at yards. However, over the years the standard shunting loco later classified as 08 performed many marshalling and shunting duties all over the Scottish rail system. With its basic design stemming from a pre-war design, various types were operated by the big four railway companies, i.e. the London Midland & Scottish, London & North Eastern, Southern and the Great Western Railway. The BR standard 0-6-0 diesel electric shunting loco, or Class 08, emerged and after nationalisation of the UK network in 1948, many 08s were produced with a full-width cab at one end. The 08s offered good all-round visibility with the driving controls on either side of the cab. Originally fitted with vacuum brakes, the 08s received many modifications over the years, including dual-brake fitment and, from the 1980s, air brakes only. With over 1,000 Class 08 locos built, they could be found all over Scotland.

How did it all begin? By the mid-1940s, in the years leading up to the start of nationalisation, considerable experience had been gained in the use of diesel shunting locos, particularly on the London, Midland and Scottish Regions of Britain. Similar examples of Class 08s were in use on most areas of the country, but by far the biggest user was the LMS, which by 1946 had the largest fleet of diesel electric locos in the world outside the USA. The various private companies within the UK soon found that well-maintained diesel shunting locos could work twenty-four hours a day as opposed to the many restrictions with their steam shunting loco counterparts. Diesel shunters could also work virtually constantly and operated for weeks at a time before being required to refuel, plus they performed well within shunting yards. The key to this success was a strict maintenance plan, which was based on a day's worked policy with maintenance and inspections carried out every 16, 32, 190 and 380 days. Larger maintenance work, such as intermediate and general works overhauls, were based on miles of operation being stipulated by BR at 30,000 and 60,000 miles. However, in reality these figures were hard to achieve as the average speed of 08s in freight yards was not more than 5 mph. As a result, the maintenance policy was changed and later based on hours worked rather than distance travelled.

In the period leading up to the formation of British Railways in 1948, several companies placed orders with their own workshops for shunting locomotives, which were to have large power outputs, all using standard designs with an engine capable of 350 hp (261 kw). Liveries of these locos took some years to complete and covered the transition period from regional businesses to when the whole system was renationalised. Early in the years of the British Transport Commision (BTC), plans were set out to design and build a standard large shunting locomotive, mainly based on the earlier LMS designs but incorporating the best features from all the pre-nationalisation companies. As the new design was required to operate unrestricted over most tracks, it was announced in the early design process that the standard engine would be the

English Electric 6K. However, engine mounting and equipment positions were to be provided for the inclusion of other manufacturers' engines and generators if required. This pleased other oil-engine builders as otherwise it was foreseen that the English Electric company would dominate the shunter market.

By 1950, the BTC announced a major shunter modernisation plan. This called for a shunter with a single cab based on the pre-war LMS design, measuring 29 feet 3 inches (8.92 metres) with a wheelbase of just 11 feet 6 inches (3.50 metres). However, to obtain improved rail clearances for branch lines, a wheel diameter of 4 feet 6 inches (1.37 metres) was stipulated, an increase of 6 inches (152.4 mm) from most of the previous shunter designs. Overall weight was also another important factor to give the best operation, and this was set at 49 tons, giving a maximum axle load of 16.33 tons with full supplies. Based on the then standard English Electric power unit and generator, a tractive effort of 35,000 lbs or 32 per cent of engine weight was attainable. The top design speed was 20 mph (32.18 km/h). At the time, in the early 1950s, all Class 08s were to be fitted with vacuum brakes and standalone air brakes.

The first order for the standard 08 shunting loco was placed in spring 1950 for twenty-five locos, with power and control equipment ordered from English Electric and mechanical assembly of the 08 shunters awarded to the former LMS works in Derby. This site was already associated with the building of the LMS 0-6-0 diesel electric shunting locos. Under the BR modernisation numbering system, locos were allocated in the 13000 series, with the first loco emerging from Derby Works on 11 October 1952. The loco was the subject of major works testing, followed by a period of operation under controlled conditions at Toton Yard, Nottinghamshire. After this was completed, No. 13000 was moved to its new home on the Western Region at Tyseley (84E). The BTC then quickly placed follow-on orders for further shunters as deliveries were required to help reduce operational costs in yards all over the country, and more orders were placed at BR workshops at Derby, Crewe, Darlington, Doncaster and Horwich. Eventually a fleet of nearly 1,200 standard shunters were built, being introduced over a ten-year period. In order to test the use of power unit/generator groups, some variations to the standard design emerged with three engine-generator pairings. The first to emerge was a Blackstone ER6T 350-hp (261-kw) engine with British Thomson Houston (BTH) electrical equipment. This combination was installed on locos 13152–13166 between February and October 1955 and built in Darlington works. The next change of equipment, and the most successful with several repeat orders, was the pairing of the Blackstone ER6T prime mover with GEC equipment. This was originally fitted to Darlington-built No. 13137 in March 1955 and it was eventually also used in batches 13137–13151, D3439–D3453, D3473–D3502, D3612–D3651 and D4049–D4094 emerging in the period from 1955–62.

The third test grouping concerned were Nos 13117–13126, which were built at Derby from 1955 and fitted with a Crossley ESNT6 engine and Crompton Parkinson electrical equipment. During the building of this batch, a total of twenty-six locomotives destined for the Southern Region of BR, were fitted with revised gearing to allow a speed increase of 7.5 mph (2 km/h) to 27 mph (43.5 km/h). This had been requested by the Southern Region authorities as the locos were to be used on a large number of local freight trips running over the main lines and it was felt that a lower speed could cause havoc with the intense EMU passenger services in that area. In terms of body design, the Southern 08 locos were more or less the same as the main orders. Some of these locos were later transferred to Scotland. On the standard 08 design, the main bay housed the generator and auxiliary generator directly in advance of the cab. The main fuel tanks were mounted on the running frame, extending in height to just below the forward-facing window cab. Ahead of the generator bay was the main engine compartment, where the engine was mounted centrally on the frame and access was from hinged side doors and a slide and removable roof section. At the nose end of the body was the cooler group/radiator with a fan on the front end. Battery boxes were mounted on the side frame, as were the vacuum exhauster and air compressors. These were all accessed by hinged flap doors, with the battery box having a hinged and removable lid.

Nose-end equipment consisted of a screw/shackle coupling with a vacuum pipe for train braking. Electric end marker lights were provided, two of which could display a red rear indication by use of a filter. When first introduced, a number of 08s emerged painted in BR locomotive black livery offset with a red buffer beam. In 1948, the 13xxx series running numbers were put on the cab sides; these were in white with a lion logo on the engine room side doors. Progressively, standard BR loco green livery was applied, with the British Railways lion holding a wheel motif on either side of the body's side doors or battery box side panel. The application of a form of warning end for yard and track staff was a major consideration to the rail industry in the 1950s. Various trial applications of white or yellow end warning panels were applied and shown to the operations staff at various locations in Scotland. One of the earliest applications of a visibility end was on 08 loco No. 13297 in February 1957, which consisted of a white end with black horizontal lines. In August 1958, a trial application of a half wasp end was applied to Nos D3682 and D3683 by BR Doncaster Works, which later became the early prototype of the standard wasp ends and were adopted on all 08s in Scotland and the UK. Progressively from 1962, yellow and black wasp ends were applied to all 08s during classified overhauls at BR workshops.

A total of 363 08 locos emerged carrying the 13xxx 1948 series numbers before the 1957 D3xxx and D4xxx series numbers were introduced. Repaints after mid-1967 received the then new BR blue livery offset by full yellow and black wasp ends, while the BR double-arrow logo and running numbers were applied. In the late 1960s, under the numeric classification system, prior to the introduction of TOPS, the standard 0-6-0 shunters were then classified as 08s. Those used on the Southern region of BR became 09s and the Class 1 identity was given to locos fitted with Blackstone/GEC equipment. Renumbering into the five-digit class-prefixed TOPS system commenced in 1973, with the majority of locos renumbered by 1975. No members of the Class 10 group survived in traffic long enough to receive TOPS numbers. Over the years, with such a large pool of locos, a number of modifications have been made. One of the most significant was the lowering of the cab heights on five locos at Landore Depot in Wales to allow operation over the restricted clearances of the Burry Port and Gwendraeth Valley line in West Wales and on some Scottish branch lines, particularly the Auchmuty branch line from Thornton Junction where No. 08515 (HA) worked Russell paper mill traffic. It also had its roof lowered to allow passage through tight bridges on this branch line.

From the 1980s, when air brakes were becoming the norm for most UK trains, the vast majority of Class 08s still in traffic were modified for dual-brake operation. This was included as part of a classified works overhaul and required major internal, external and cab work modifications. Modified locos could be identified by the extra frame mounted on the equipment box housing the air compressor and buffer beam mounted main reservoir and brake pipes. A large number of 08s were also fitted with extra high-level air pipes, allowing easy brake pipe connections to waist level for easier connections to the brake pipes fitted on multiple unit stock. On these locos the nose-mounted air connections were operated by a dual-valve handle operating both brake pipe and main reservoir connections. In subsequent years, with the reduction in the number of vacuum-fitted wagons and coaching stock, it was agreed to adopt a policy of removing the vacuum brake equipment and rendering most 08 locos to be air brake only. This modification was clearly identified by the omission of the vacuum pipe from the buffer beam.

In more recent years headlights have become a standard requirement. A large number of 08 locos have been fitted with various designs of high-intensity headlights. Although originally introduced with just a conventional screw coupling, as more advanced coaching stock and wagons were introduced, particularly MK3 coaching stock, it became obvious that a modification would have to be made to the Class 08 fleet. This problem was overcome by the fitting of a basic drophead knuckle coupler compatible with HST buckeye couplers. Having a drophead, the 08s could then be used for normal shunting or the knuckle lifted and locked into place to attach to HST stock. In more recent years, some local modifications to couplings have been made, including variable height loop couplers on the Railcare-owned 08s at Wolverton Works in Buckinghamshire. With no obvious replacement for the standard 08s in the 1990s, it was authorised to put a number of the class through a major works refurbishment programme and an upgraded top speed of 27 mph (43.5 km/h) made as standard. Originally this project was for a significant number of 08 locos to be upgraded, and was announced by the British Railways Board (BRB). However, in reality, the work was sub-contracted to RFS Engineering of Kilnhurst, near Rotherham, and in fact only actually involved a small number of Class 08 locos.

With the reducing need for Class 08 shunting power throughout the whole of the UK, mostly down to the closure of freight yards and the use of train locos for shunting duties, the need for the standard 08 shunting loco hugely declined from the 1990s. In the period leading up to rail privatisation, and from 1996, huge cutbacks were made into the 08 fleet requirement. By 1998, only around 500 Class 08s remained in the UK, with approximately fifty left in Scotland. The sectorisation within the freight sub-sector brought further changes to the allocations and redeployment of the 08s, with all of the three major freight companies receiving an allocation of 08s to their fleets. At the same time, some of the new private freight companies also purchased 08 locos and most of the passenger operators were also allocated a small number for depot shunting and piloting operations. Today, Class 08s are operated by DB Cargo, Freightliner, Direct Rail Services, GB Railfreight, First Great Western, East Midland Trains, East Coast, Transport for London, Southern, London Midland, Eurostar UK, DB Arriva, RVEL, Serco, Foster Yeoman, West Coast Railways, Alstom, Railcare, NRM York, Wabtec, RML, Hunslet Engine Company, HRNC, Peterbrook and Advanti West Coast. British American Railway Company also offers a lease and hire service. Following their useful life working the nationalised UK network and the subsequent private sector, a significant number were sold for industrial use. They operated virtually all types of industrial settings, ranging from quarries and steelworks to paper mills and food-processing plants. With a surplus of Class 08 locos, most private companies could see the financial benefits of buying their own Class 08s and employing a local member of staff to drive them, rather than hiring them with a driver from a mainstream operator. The preservation sector has also found homes for a large number of 08 diesel shunters, either powering works trains or passenger services. In the main these have been fully restored to full working condition, many reflecting true liveries and equipment positions allowing more of our history to be recreated for future generations to enjoy.

In 1964, a need was foreseen for some more powerful shunting locos to operate the hump yard at Tinsley Yard, near Sheffield. One standard 08 was deemed insufficient to provide the required traction and braking power for propelling heavy trains over the humps to gravitate into the yard sidings. It was agreed to rebuild three pairs of standard 08s together, remove the cab and driving equipment from one and fix extra weight to both, thereby improving adhesion. This was achieved by attaching plates to the buffer beams. Control over both locos was placed with the driver of the master unit, including the starting and stopping of the slave 08 loco and fuel transfer. The conversion contract for the three selected 08s was awarded to Darlington

Works and conversion of six 08s was undertaken from the summer of 1965. These 08s were semi-permanently coupled cab to cab and painted in BR green with wasp ends. After a period of operation, these 08 locos were re-marshalled to operate elephant style, with the nose of the master 08 coupled to the cab position of the slave 08 loco. In the late 1960s and early 1970s all three modified 08s were redesignated as Class 13s and were repainted into standard BR blue livery. This trio of special Class 08s incorporated special in-cab signalling and radio equipment for use on the Tinsley Yard humps. As traffic began to reduce from the mid-1980s, the hump yard at Tinsley eventually closed and the three special Class 13s were withdrawn from service; sadly, all were broken up. None of these super 08s were transferred to Scotland, although consideration had been given to moving some to Mossend Yard for shunting operations near Bellshill.

In 1985, three locomotives were reduced in height for use on the Burry Port and Gwendraeth Valley Railway in south-west Wales and became Class 08/09s. In Scotland, No. 08515 also had its height reduced to allow it to work freight workings to and from Auchmuty from Thornton Yard until it closed in the early 1990s. The remainder of the class were reclassified as sub-Class 08/9s. A further two were converted to 08/9s in 1987. The first locomotive to be withdrawn was D3193 in 1967. Four other machines were withdrawn before TOPS reclassification in 1973. Withdrawals continued in subsequent decades until, by the beginning of the 1990s, most of the Class 08/1s had been withdrawn. As part of the privatisation of British Rail in the mid-1990s, most of the survivors passed to English, Scottish & Welsh Railways (EWS), with some going to passenger operators for use as depot shunters. At the same time as the withdrawals many were also purchased by heritage railways.

By mid-2008, EWS had over forty Class 08s in operation, with a greater number stored. Freightliner also had about five in operation, as did locomotive company Wabtec Rail, and First Group operated fewer than five. Some of the original 08s found work at industrial locations – two for Foster Yeoman, one for Mendip Rail, one for Corus, one at ICI Wilton and two went to the English China Clay Company amongst others. A few others found their way into smaller railway-related businesses. Sixteen English Electric 060DE 350-hp locomotives based on the Class 11/Class 08 design but modified for 5-foot 3-inch gauge (1,600 mm) were built new and exported between 1951 and 1953 to Australia, entering service on the Victoria Railway as the F Class. Five examples of the Class 08 were exported to Liberia – Nos 3047, 3092, 3094, 3098 and 3100. All five locos remain in Liberia and have been considerably robbed of parts and are unlikely see service in the future.

In 2007, Nos 08738 and 08939 were equipped for multiple operation at Toton TMD Nottingham and repainted into Euro Cargo Rail livery before being sent to France in April 2009 for work on engineering trains in connection with new high-speed lines being built from London to Dover. As the standard general-purpose diesel shunter on BR, almost any duty requiring shunting would involve a Class 08, thus the many locations where two portions of a train were merged, or where additional stock was added to a train, were hauled briefly by a Class 08. As such, the class was a familiar sight at many major stations and terminals. Certain Class 08 locomotives were modified from the standard class by being given headlights and cut-down bodywork – the overall height was reduced to 11 feet 10 inches (3.61 metres) – for use on the Burry Port and Gwendraeth Valley Railway up to Cwm Mawr. In 2007, three were also later used on infrastructure trains on the Manchester Metrolink construction work. The Class 08 locomotives were also the basis for the character Diesel in the railway series of books written by Revd W. Awdry and the subsequent *Thomas & Friends* television series. Class 08s have also provided the basis for several other characters in the series, most notably 'Arry and Bert, Splatter and Dodge (Splodge), and Paxton and Sidney.

Depot Codes

AB	Aberdeen Ferryhill
AY	Ayr
CD	Crewe Diesel Depot
ED	Eastfield
DE	Dundee
GM	Grangemouth
HA	Haymarket
HQ	Headquarters (BR)
IS	Inverness
KD	Carlisle Kingmoor
ML	Motherwell
PO	Polmadie
PT	Perth
TJ	Thornton Junction
ZA	RTC Derby

List of Preserved British Rail Class 08 Locomotives in 2023

Numbers Carried (current in bold)				**Name**	**Location**	**Current Status**	**Livery**
13000	D3000	–	**13046**	–	Peak Rail	Awaiting overhaul; first one built	BR Black
13002	D3002	–	–	–	Plym Valley Railway	Operational	BR Black
13014	**D3014**	–	–	*Samson*	Dartmouth Steam Railway	Operational	BR Green, Dart Rail insignia
13018	**D3018**	08011	–	*Haversham*	Chinnor and Princes Risborough Railway	Operational	BR Green
13019	**D3019**	–	–	–	Cambrian Heritage Railways	Operational	Blue and White
13022	**D3022**	08015	–	–	Severn Valley Railway	Operational	BR Green
13023	D3023	**08016**	–	*Geoff L Wright* (not carried)	Peak Rail	Operational Last of first Batch	BR Blue
13029	D3029	08021	–	–	Tyseley Locomotive Works	Operational	BR Black

Numbers Carried (current in bold)				Name	Location	Current Status	Livery
13030	D3030	08022	–	*Lion*	Cholsey and Wallingford Railway	Operational	Guinness Black
13044	D3044	**08032**	33	*Mendip*	Mid Hants Railway	Undergoing repairs	Yeoman Blue
13059	**D3059**	08046	–	*Brechin City*	Caledonian Railway (Brechin)	Operational	BR Green
13067	D3067	**08054**	–	–	Embsay and Bolton Abbey Railway	stored awaiting re-wiring	BR Blue
13074	D3074	08060	–	*Unicorn*	Cholsey and Wallingford Railway	Operational	Guinness Black
13079	D3079	08064	–	–	National Railway Museum	Operational	Black
13101	D3101	–	**3101**	–	Great Central Railway	Operational	BR Green
13167	**D3167**	08102	–	–	Lincolnshire Wolds Railway	Operational	BR Green
11374	D3174	08108	**3174**	*Dover Castle*	Kent and East Sussex Railway	Under overhaul	Black
13180	D3180	08114	–	–	Great Central Railway (Nottingham)	Undergoing restoration	BR Green
13190	**D3190**	08123	–	*George Mason*	Cholsey and Wallingford Railway	Operational	BR Green
13201	**D3201**	08133	–	–	Severn Valley Railway	Under Overhaul	BR Green
13232	D3232	**08164**	–	*Prudence*	East Lancashire Railway	Operational	BR Blue
13236	D3236	08168	–	–	Battlefield Line Railway	Operational; on loan to the Bluebell Railway	BR Black
13255	D3255	–	–	–	Private site, Leavening, North Yorkshire	To be restored	N/A

Numbers Carried (current in bold)				Name	Location	Current Status	Livery
13261	D3261	–	–	–	Swindon and Cricklade Railway	Operational	BR Green
13265	D3265	08195	WD40	*Mark*	Llangollen Railway	Operational	BR Black
13272	D3272	**08202**	–	–	Avon Valley Railway	Operational	BR Blue
13290	D3290	**08220**	–	–	Great Central Railway (Nottingham)	Operational	BR Blue
13308	D3308	**08238**	–	*Charlie*	Dean Forest Railway	Light duties only	BR Blue with Wasp Stripes
13336	D3336	**08266**	–	–	Keighley and Worth Valley Railway	Operational	BR Green
13358	**D3358**	08288	–	–	Mid Hants Railway	Awaiting repair	BR Black
–	D3401	**08331**	–	–	Midland Railway - Butterley	Operational	Black
–	D3420	08350	–	–	Crewe Heritage Centre	Scrapped. Early 2009	BR Green
–	D3429	**08359**	–	–	Chasewater Railway	Operational	BR Blue
–	**D3462**	08377	–	–	Mid Hants Railway	Operational	BR Green
–	**D3411**	08411	–	–	Colne Valley Railway	under restoration	BR Dutch grey
–	D3503	**08388**	–	–	Private Site	Undergoing restoration	Trainload Petroleum
–	D3551	**08436**	–	–	Swanage Railway	Operational	Black
–	**D3558**	08443	–	–	Bo'ness and Kinneil Railway	Operational	BR Green (No Logo)
–	D3559	**08444**	–	–	Bodmin and Wenford Railway	Operational	BR Green

Numbers Carried (current in bold)				Name	Location	Current Status	Livery
–	D3560	**08445**	–	–	Daventry IRFT	Awaiting repair	Yellow
–	**D3586**	08471	–	–	Severn Valley Railway	Operational	BR Green
–	D3588	**08473**	–	–	Dean Forest Railway	Scrapped; source of spares	BR Blue
–	**D3591**	08476	–	–	Swanage Railway	Operational	BR Green
13594	D3594	08479	–	–	East Lancashire Railway	Operational	BR Black
–	D3596	**08481**	–	–	Vale of Glamorgan Railway	Scrapped EMR	BR Blue
–	**D3605**	08490	–	–	Strathspey Railway	Operational	BR Green
–	D3610	**08495**	–	–	North Yorkshire Moors Railway	Operational	BR Blue
–	D3690	**08528**	–	–	Great Central Railway	Operational	BR Green
–	D3723	**08556**	–	–	North Yorkshire Moors Railway	Operational	BR Green
–	**08576**	D3743	–	–	??	Awaiting restoration	BR Blue
–	D3757	**08590**	–	*Red Lion*	Midland Railway – Butterley	Operational	BR Blue
–	D3759	08592	**08993**	–	Keighley and Worth Valley Railway	Operational	EWS Maroon and Gold
–	D3761	**08594**	–	–	Private Site	Awaiting overhaul	BR Blue
–	D3771	08604	**604**	*Phantom*	Didcot Railway Centre	Operational	BR Green

Numbers Carried (current in bold)				Name	Location	Current Status	Livery
–	D3795	**08628**	–	–	Private Site	Awaiting overhaul	BR Blue
–	-	**08631**	–	*Betsy*	Bombardier Transportation	Operational	BR Blue
–	**D3800**	08633	–		Churnet Valley Railway	Operational	BR Green
–	D3830	**08663**	–	*St Silas* (nameplates removed)	Avon Valley Railway	Not operational at present	BR Blue
–	–	**08682**	–	*Lionheart*	Bombardier Transportation	Operational	Designed by the winner of a children's painting competition
–	D3861	**08694**	–	–	Great Central Railway	Operational	EWS
–	D3867	**08700**	–	–	East Lancashire Railway	Operational	BR Blue
–	**D3871**	08704	–	–	Private Site	Operational	BR Green
–	D3902	**08734**	–	–	Dean Forest Railway	Undergoing restoration with parts from 08473	Blue
–	**D3918**	08750	–	–	Private Site	Operational	Black
	D3926	08757			Telford Steam Railway	Operational	RES
–	**D3935**	08767	–	–	North Norfolk Railway	Undergoing overhaul	BR Green
–	**D3937**	08769		*Gladys*	Dean Forest Railway	Operational	BR Green
–	**D3940**	08772	–	*Tiny/Camulodunum*	North Norfolk Railway	Operational	BR Green
–	D3941	**08773**	–	–	Embsay and Bolton Abbey Steam Railway	Operational	BR Blue
–	D3948	**08780**	–	*Fred*	Jeremy Hosking	Operational	Dark Blue
–	D3952	08784			Great Central North		

Numbers Carried (current in bold)				Name	Location	Current Status	Livery
–	D3991	**08823**	–	–	LH Barton	Operational	Hunslet Orange and Blue
–	D3993	**08825**	–	–	Chinnor and Princes Risborough Railway	Operational	BR Blue
–	D4018	**08850**	4018	–	North Yorkshire Moors Railway	Operational	BR Blue
–	D4033	08865			East Kent Light Railway	Operational	EWS
–	D4118	08888	**08888**	–	Kent and East Sussex Railway	Operational	BR Green
–	**D4137**	08907			Great Central Railway	Operational	BR Green
–	D4141	**08911**	–	–	National Railway Museum - Shildon.	Operational	BR Blue (with small NRM logo)
–	D4145	**08915**	–	–	North Tyneside Steam Railway	Operational	BR Blue
–	D4157	08922			Great Central North		Rail Freight Grey
–	D4167	08937	–	*Bluebell Mel*	Dartmoor Railway	Operational	BR Green
–	D4174	**08944**	–	–	East Lancashire Railway	Awaiting repair	Black
–	D3854	**08995**	08687	*Kidwelly*	Shillingstone Station Project	Under Repair	EWS Maroon and Gold

Current Operations in Scotland as of 2023

One 08 based at Inverness is for shunting coaching stock for the Caledonian Sleepers. The main duty is to haul the morning arrival from Euston out of the station. Drivers for this shunting loco are provided by GBRF. A second 08 is used by ScotRail for shunting HST trailer coaches at Inverness as well. Since the introduction of HSTs to ScotRail services in 2017, Haymarket Depot and Cadder Yard now also have Class 08s to assist with shunting operations.

At Polmadie, Class 08s are used for shunting Caledonian Sleeper stock and also for moving internal coaching stock within the depot.

DB Schenkers use of Class 08s at Mossend Yard ended in October 2016. The final loco was 08632. It remained at Mossend until sold in January 2017.

Number	Livery & Remarks	Location	Operator
08754	Blue, with grey cabsides. RMS logos	Inverness Depot	Caledonian Sleepers
08648	Black with RMS logos.	Inverness Depot	ScotRail
08472	Black with Wabtec logo	Edinburgh Craigentinny	East Coast
08596	Black with Wabtec logo	Edinburgh Craigentinny	East Coast
08764	BR Blue, unbranded	Polmadie	West Coast Traincare
08887	BR Blue, unbranded	Polmadie (Down Side)	West Coast Traincare
08954	BR Blue, unbranded	Polmadie	West Coast Traincare
08447	Lilac with Russell logos	Hillington (Deanside). OOU.	Russell Group

There are three Class 08s on preserved railways:

D3059 / 08046 / Caledonian Railway, Brechin
D3558 / 08443 / Bo'ness and Kinneil Railway
D3605 / 08490 / Strathspey Railway

Two of the earlier LMS design BR Class 11 locos are also preserved in Scotland:

12052 / Caledonian Railway, Brechin
12093 / Caledonian Railway, Brechin

Scottish Class 08s

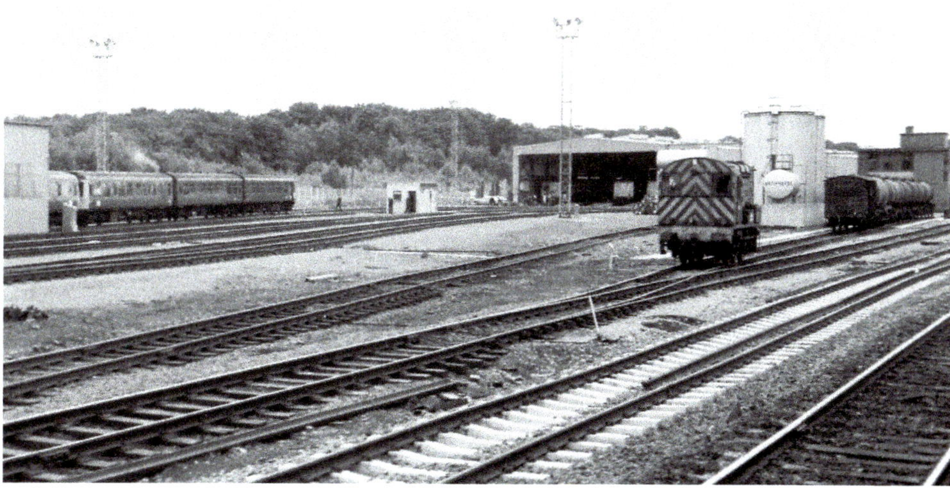

Above: No. 08754 (ED) at Corkerhill Depot, Glasgow, in-between shunting duties with DMU classes 101–126. Taken July 1979.

Below: No. 08343 (AY) approaches Byrehill Junction south of Kilwinning with a local trip from ICI Ardeer to Falkland Yard. Taken March 1979.

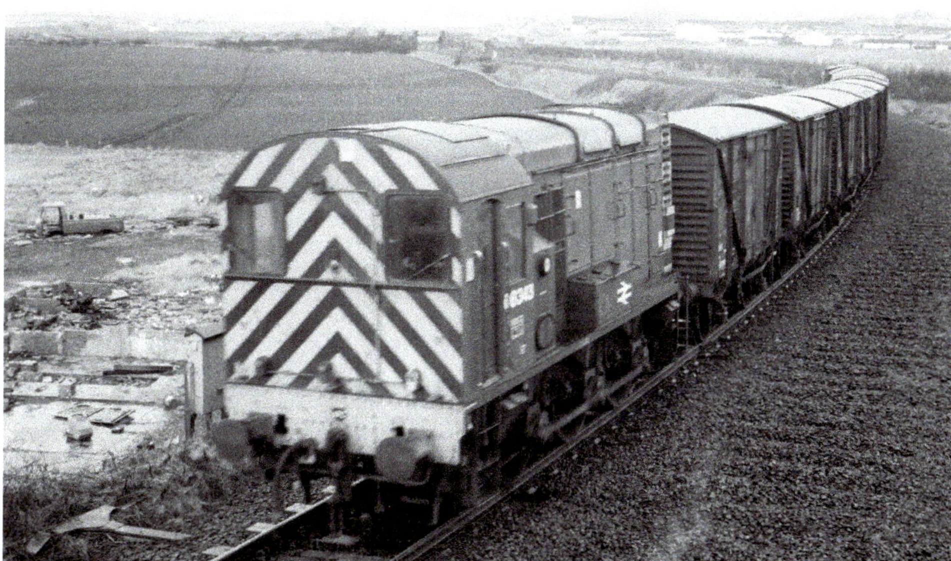

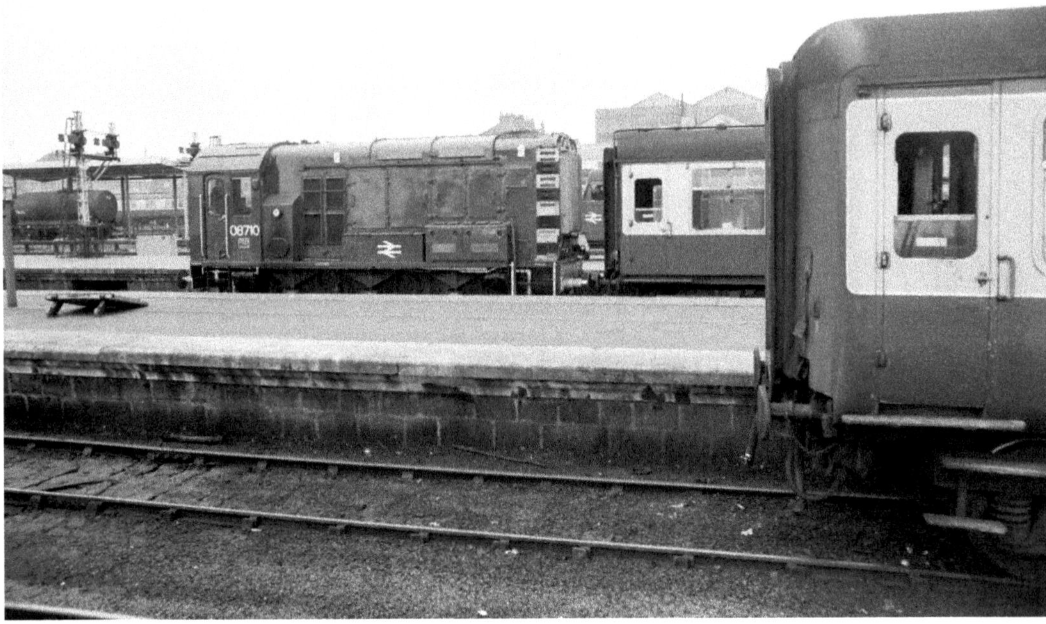

No. 08710 (AB) at Aberdeen station ready to take ECS to the nearby Clayhills Depot for servicing. Taken August 1980.

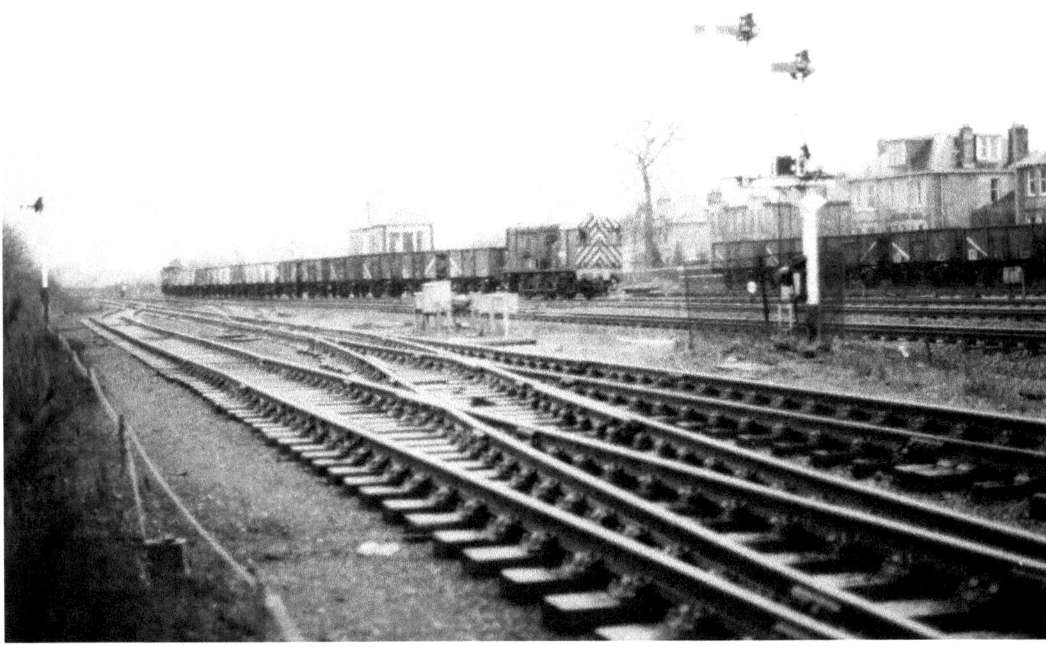

No. 08727 (ED) at Elderslie No. 1 signal box with a local coal trip working from Paisley Canal Coal Yard. Taken November 1979.

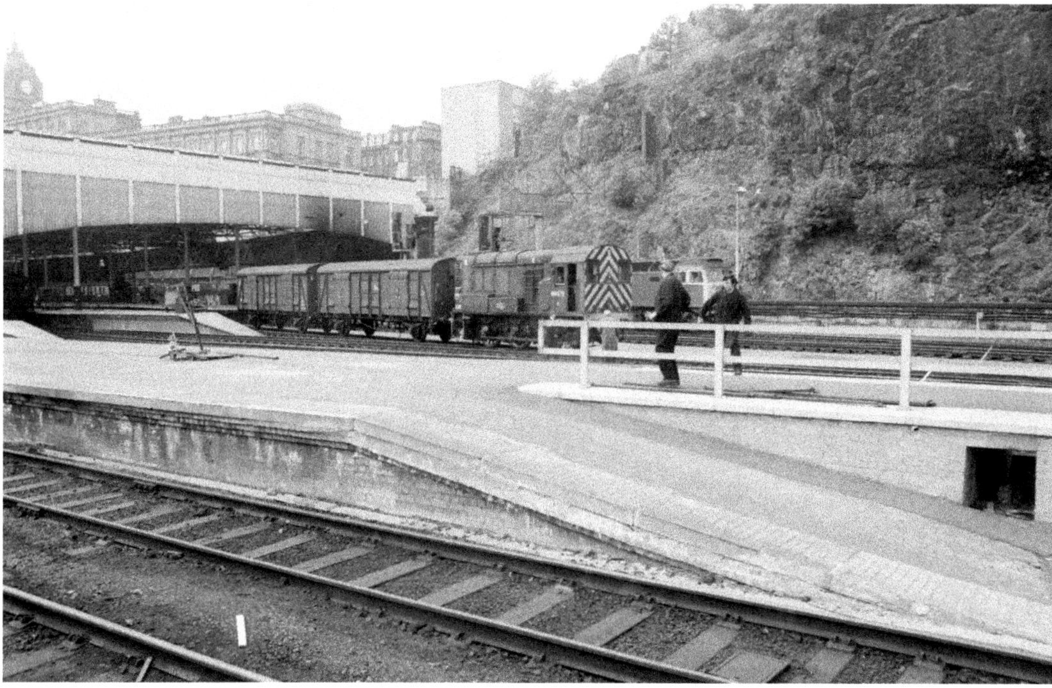

No. 08571 (HA) at Edinburgh Waverley shunting parcel vans. Taken June 1980.

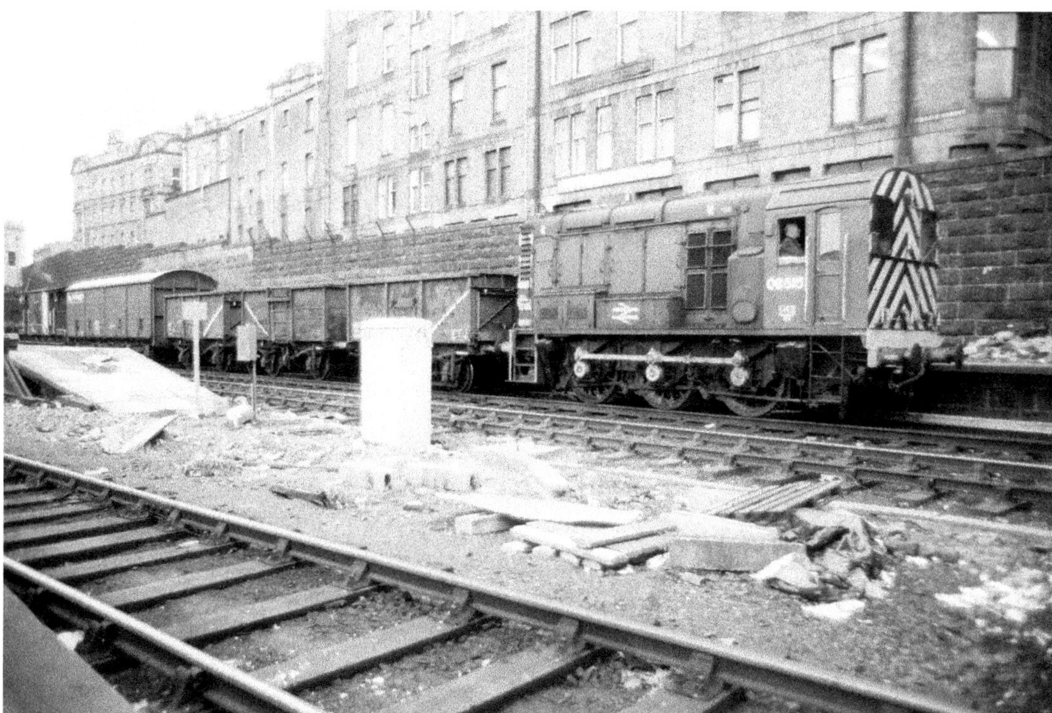

No. 08515 (AB) at Aberdeen station with a local trip to Kittybrewster. Taken August 1980.

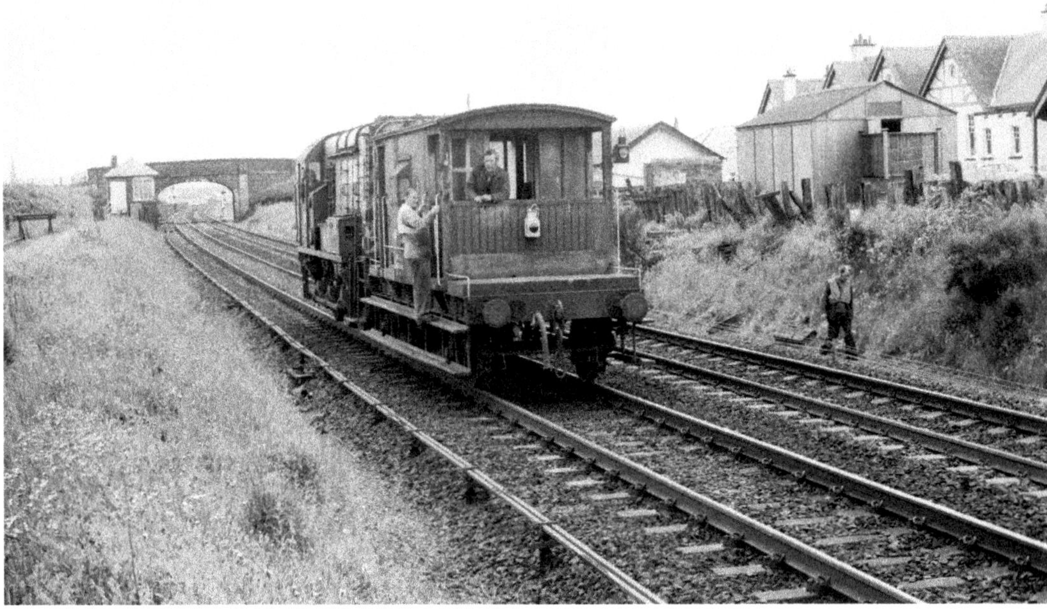

No. 08731 (ED) with brake van at Bogside, north of Irvine. I managed to get a ride on this back to Irvine station free of charge! Taken June 1980.

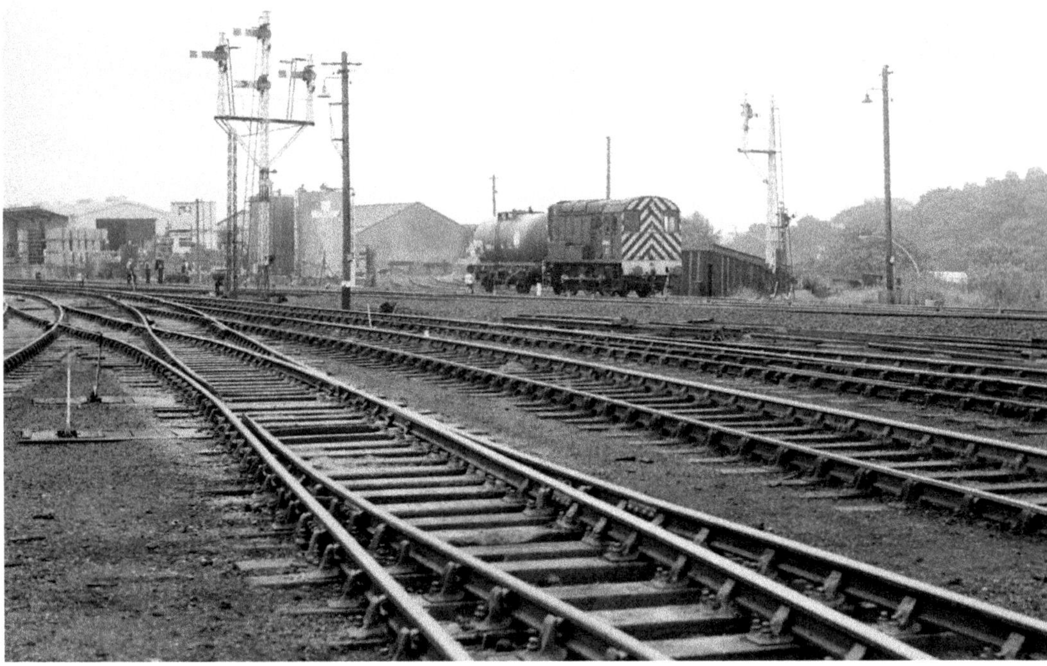

An unidentified 08 with an ESSO tank wagon at Ferryhill Depot, Aberdeen. This wagon would probably supply diesel fuel for main line locos off Inverness, Edinburgh and Glasgow Queen Street services. Taken August 1980.

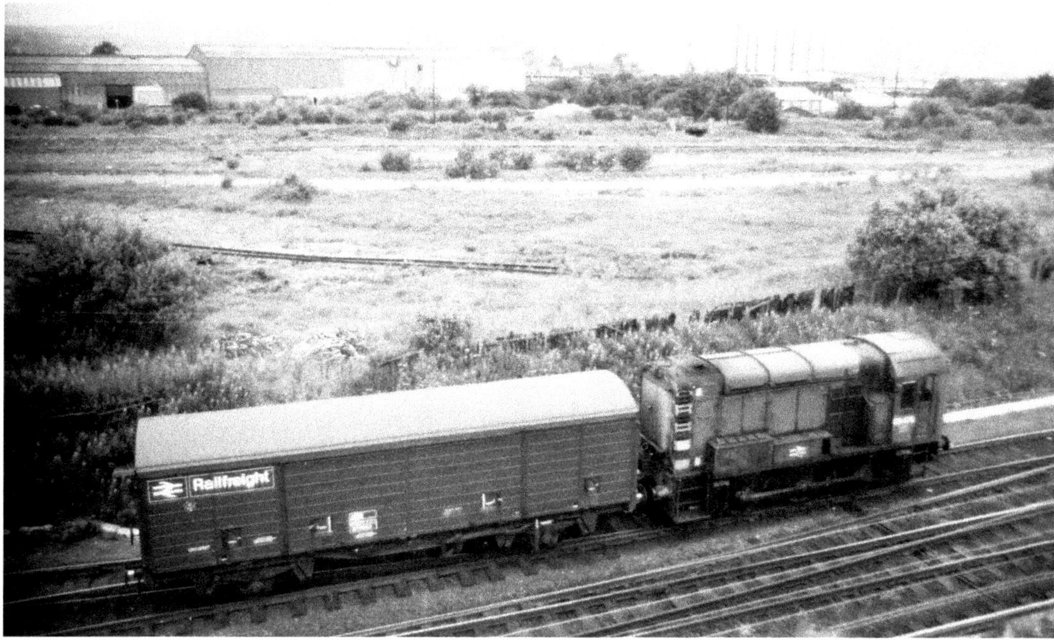

No. 08476 (AY) shunts a Railfreight van for McKinnons Road Transport Shed, situated next to Glengarnock station. Note the steelworks in the background, which was closed in 1985. Taken August 1979.

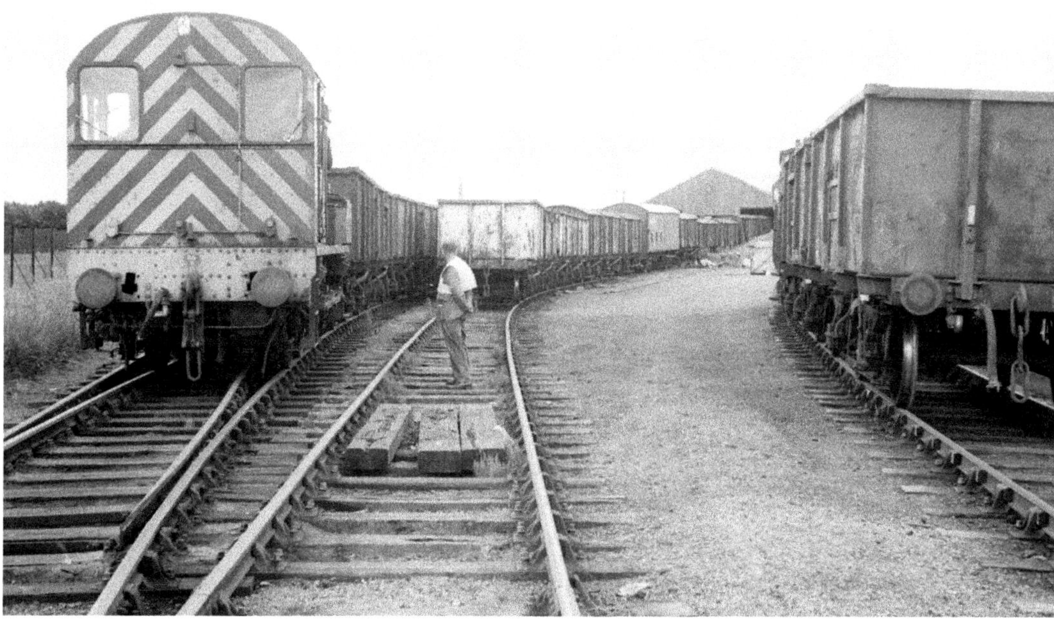

No. 08731 (AY) at ICI Snodgrass, between Irvine and Kilwinning, carrying out shunting. Note the empty coal wagons, which were mandatory, being marshalled between the loco and gunpowder vans. Taken June 1980.

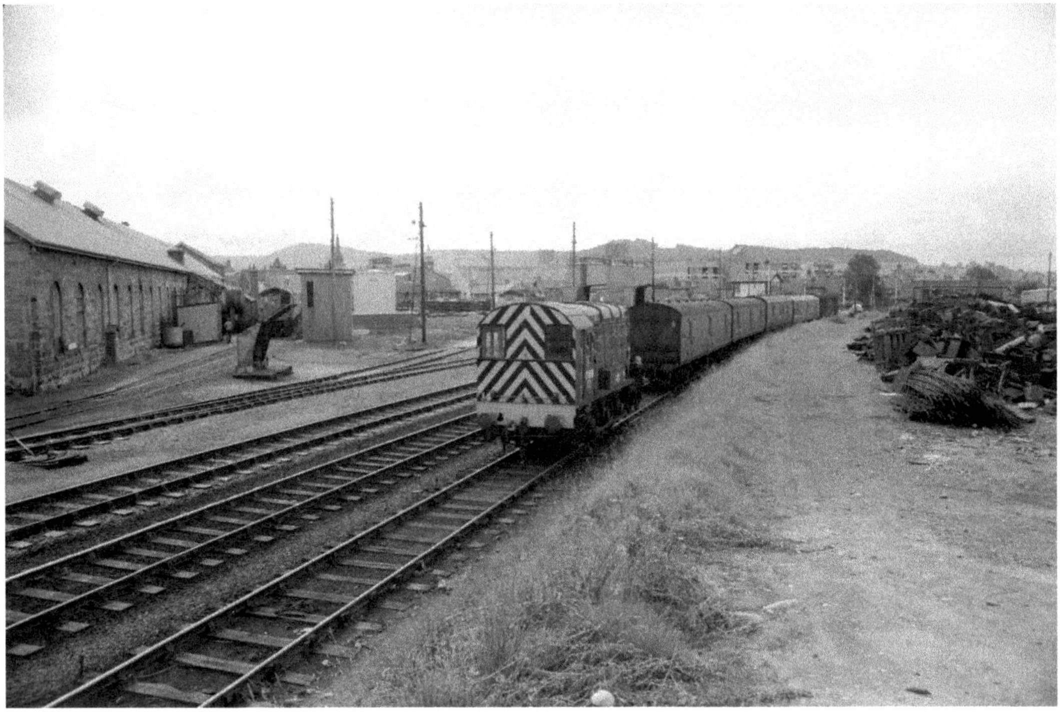

No. 08568 (IS) at Rose Street, Inverness, ready to attach to a set of newspaper coaches. This loco was sold to Railcare Ltd at Springburn Works in Glasgow in December 2000. Taken August 1980.

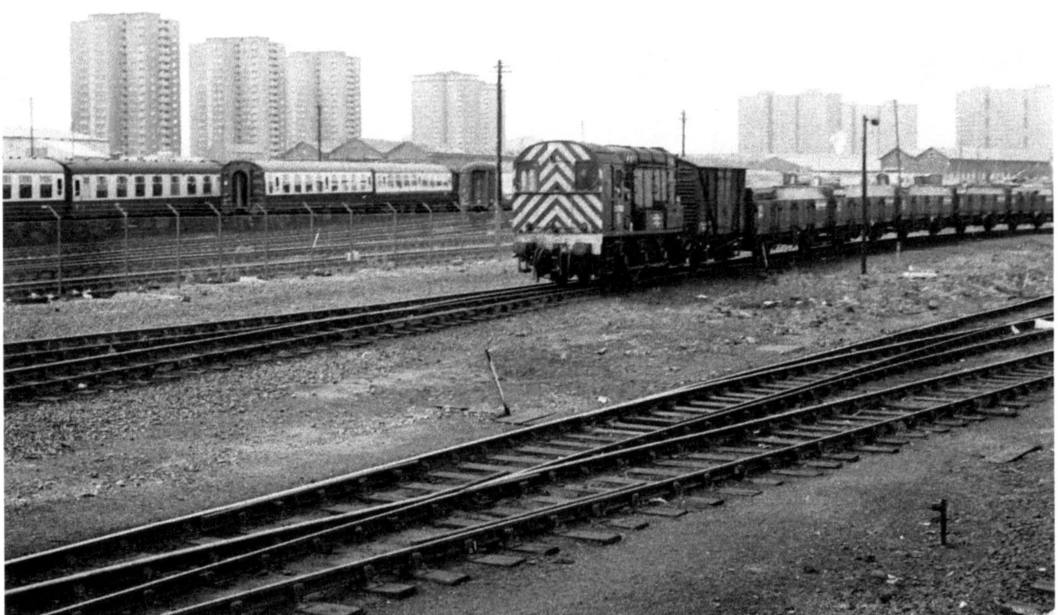

An unidentified 08 shunts wagons at Springburn Works, Glasgow. Taken July 1979.

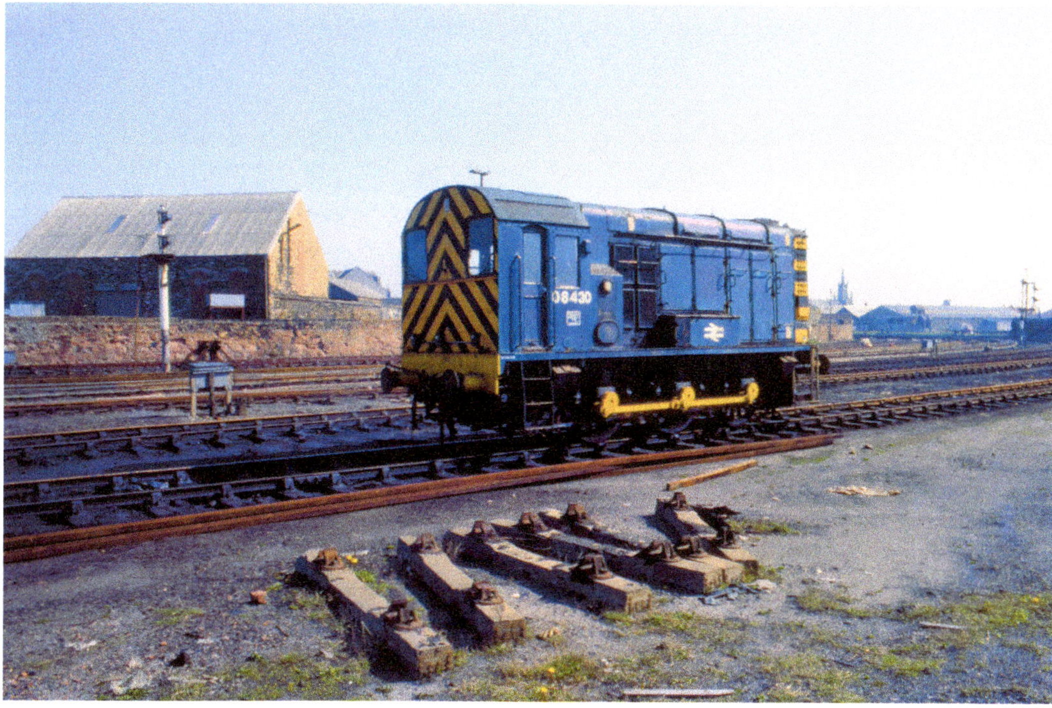

No. 08430 (AY) stabled in-between duties at Ardrossan Harbour. This loco was withdrawn from traffic in November 1985 and cut up by March 1986. Taken April 1984.

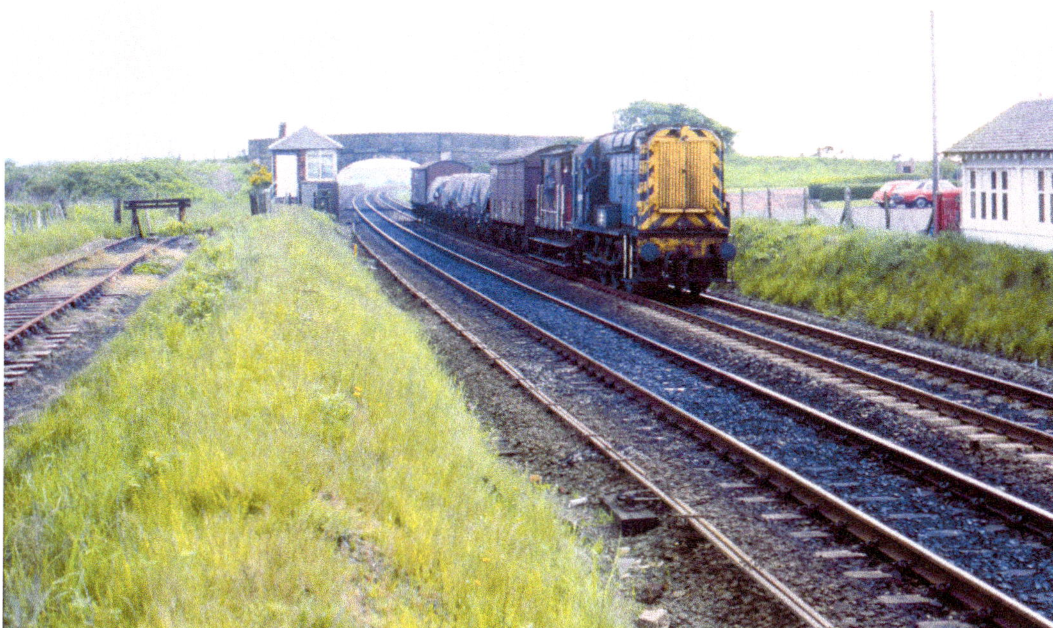

No. 08446 (AY) at Bogside, north of Irvine, with a northbound local trip working from Falkland Yard to ICI Ardeer, Snodgrass section. Taken May 1983.

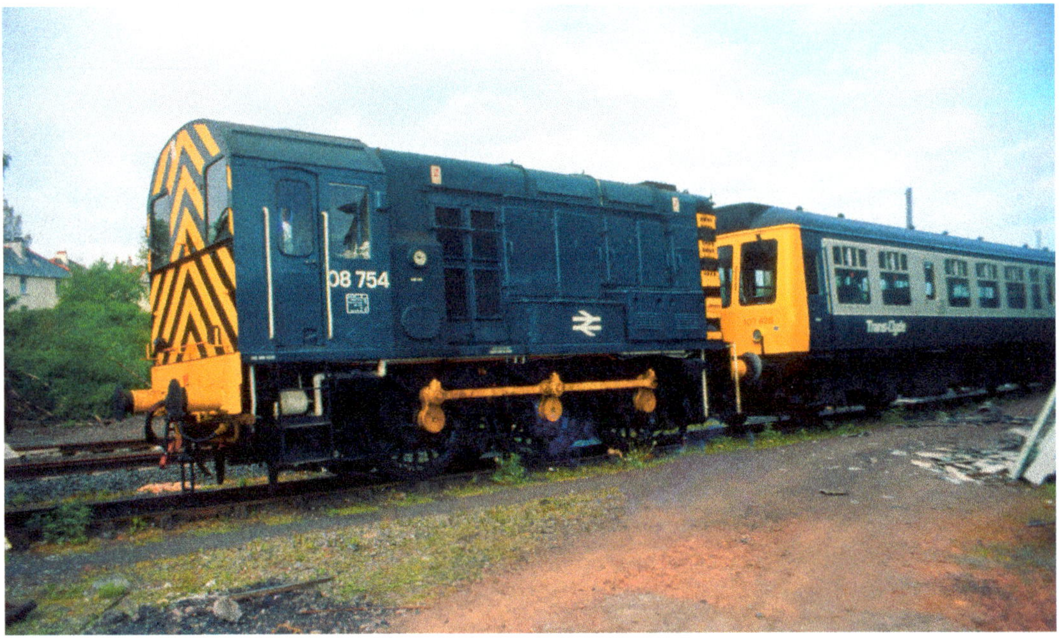

No. 08754 (ED) at Corkerhill Depot, Glasgow, shunting a Class 107 DMU power car. This loco was withdrawn from traffic in December 2000. Taken May 1984.

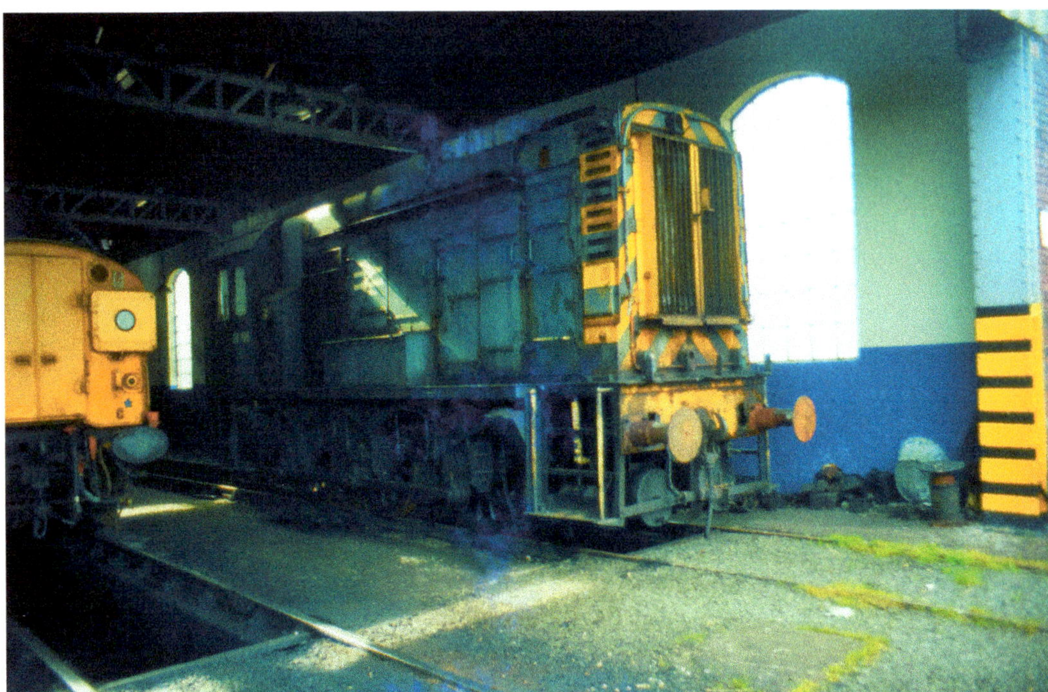

No. 08196 (GM) stabled at its home depot of Grangemouth in-between duties. This loco was withdrawn from traffic in December 1985 and was cut up for scrap at Glasgow London Goods Depot by April 1986. Taken May 1984.

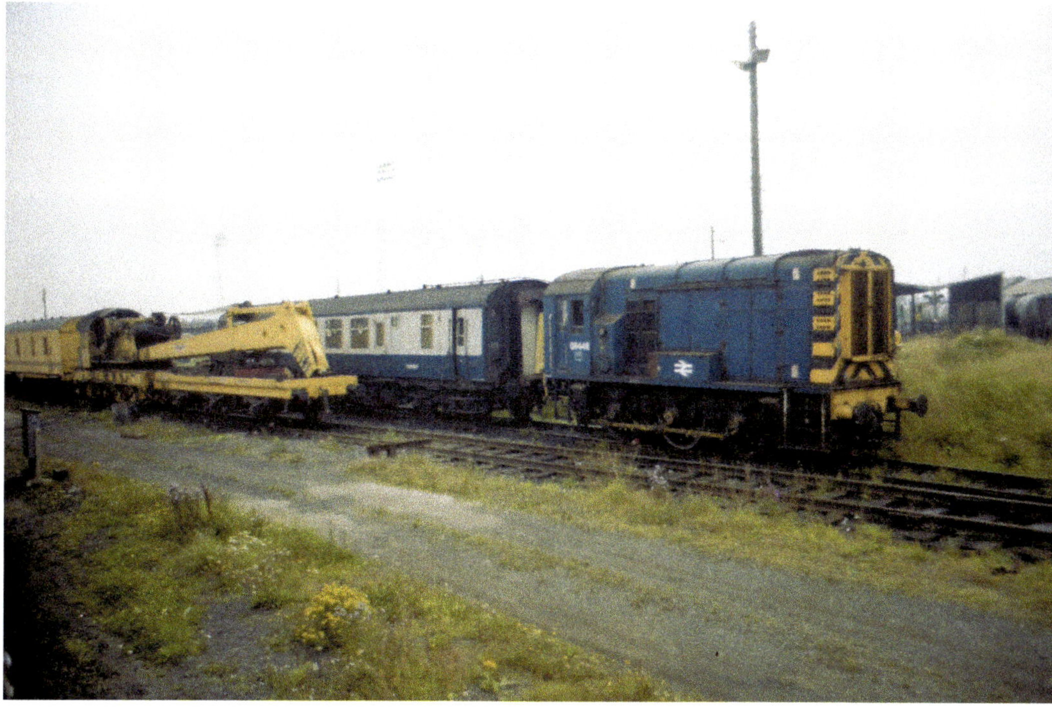

Above: No. 08446 (AY) at Ayr Depot with Class 126 DMU power car No. 51048. Taken August 1981.

Right: Two Class 08s in the snow at Kilwinning station heading from Ayr Depot to Glengarnock and Ardrossan Harbour. These locos were returning from Ayr Depot after maintenance and refuelling. Taken February 1980.

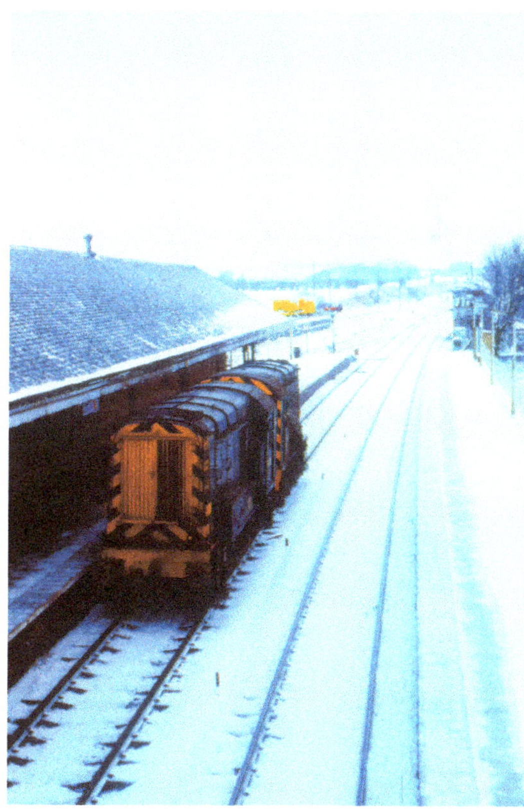

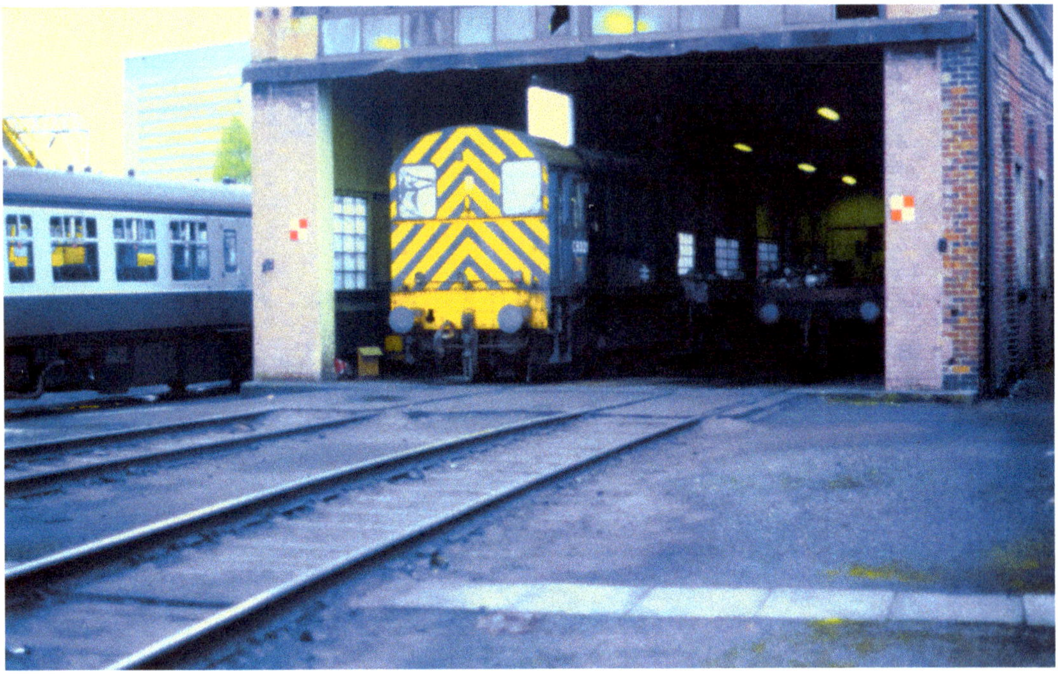

No. 08321 (ML) at Hamilton Depot stabled in-between duties. Most of these duties involved shunting DMU stock. Hamilton Depot closed in June 1982. Taken April 1982.

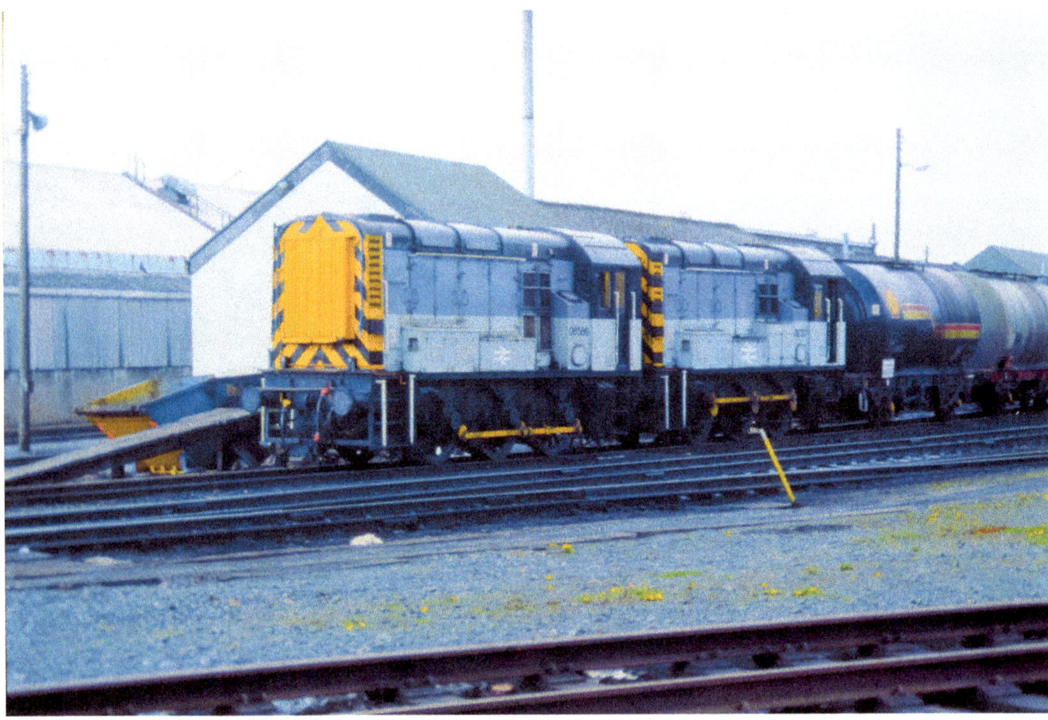

Nos 08586 and 08727 (both AY) at the fuelling road in Ayr Depot. Taken May 1991.

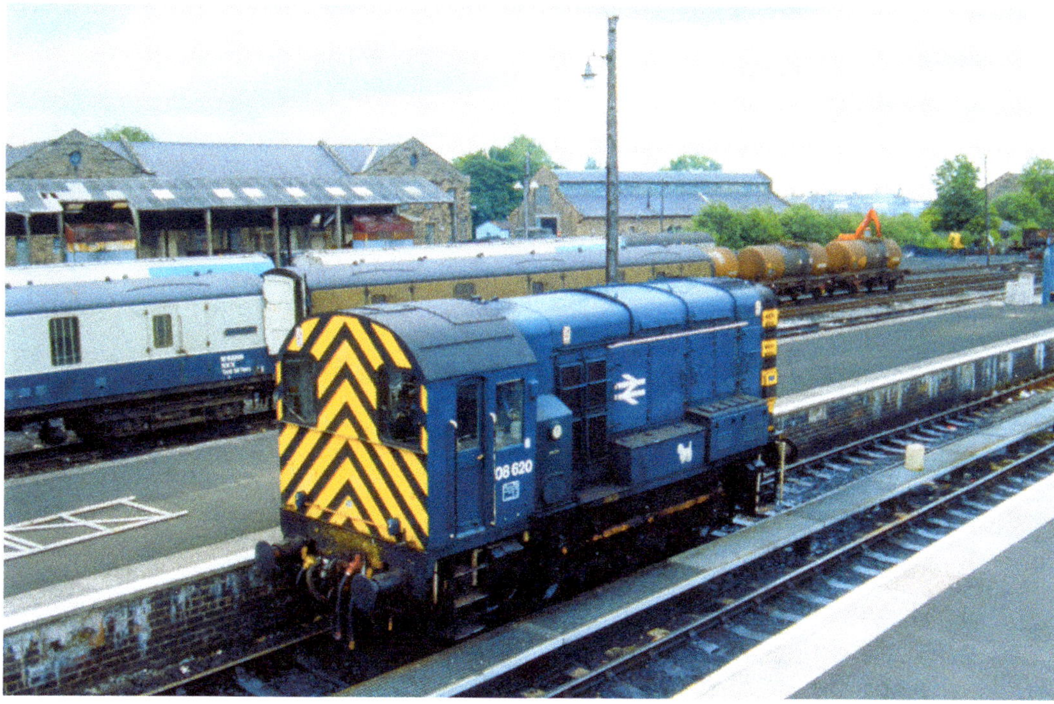

No. 08620 (ML) at Stirling station stabled between Motorail shunting duties. Taken June 1986.

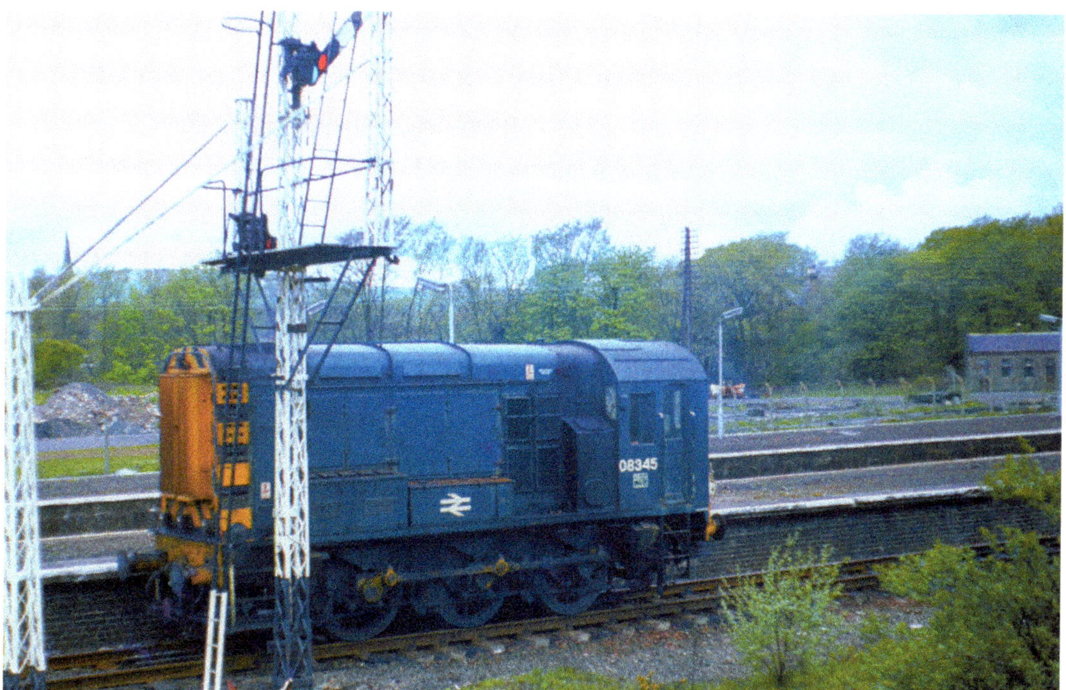

No. 08345 (AY) on the Up Slow line near Dalry en route from British Steel, Glengarnock, to Ayr Depot for refuelling and maintenance. Taken May 1978.

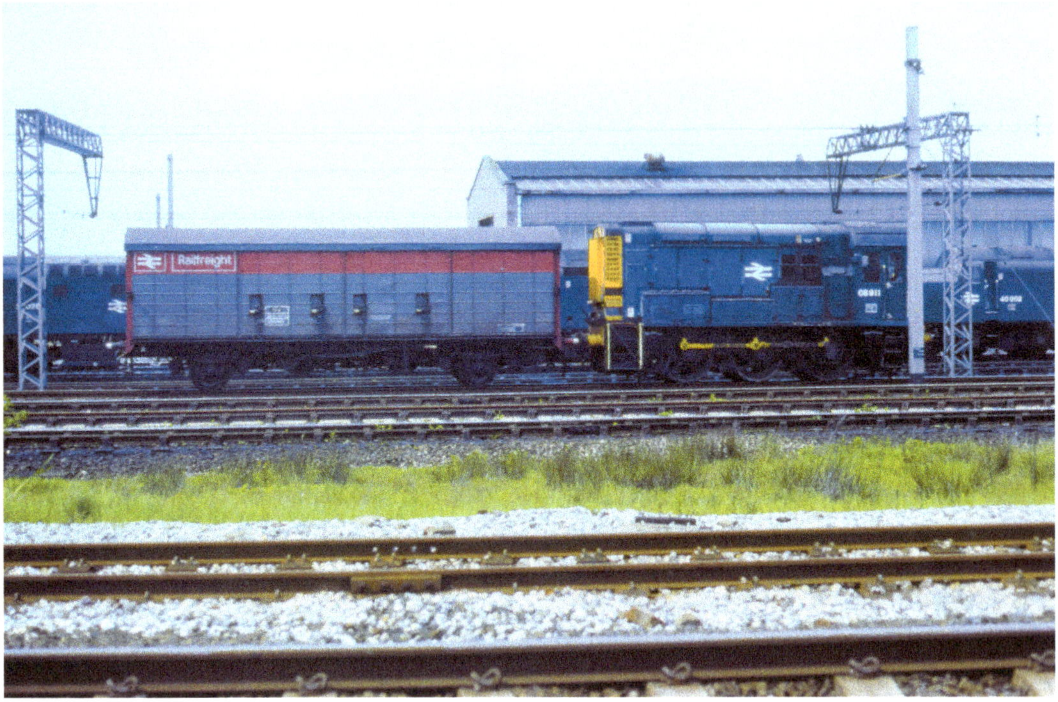

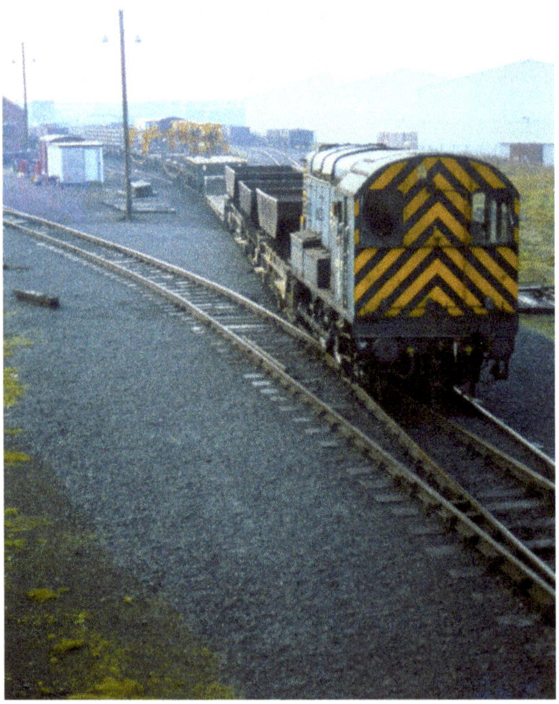

Above: No. 08911 (KD) at Kingmoor Yard shunting ZHA wagon 200481, plus a Class 40 in the background. Taken May 1983.

Left: Class 08 shunting engineers' wagons at Irvine West Yard. Taken September 1980.

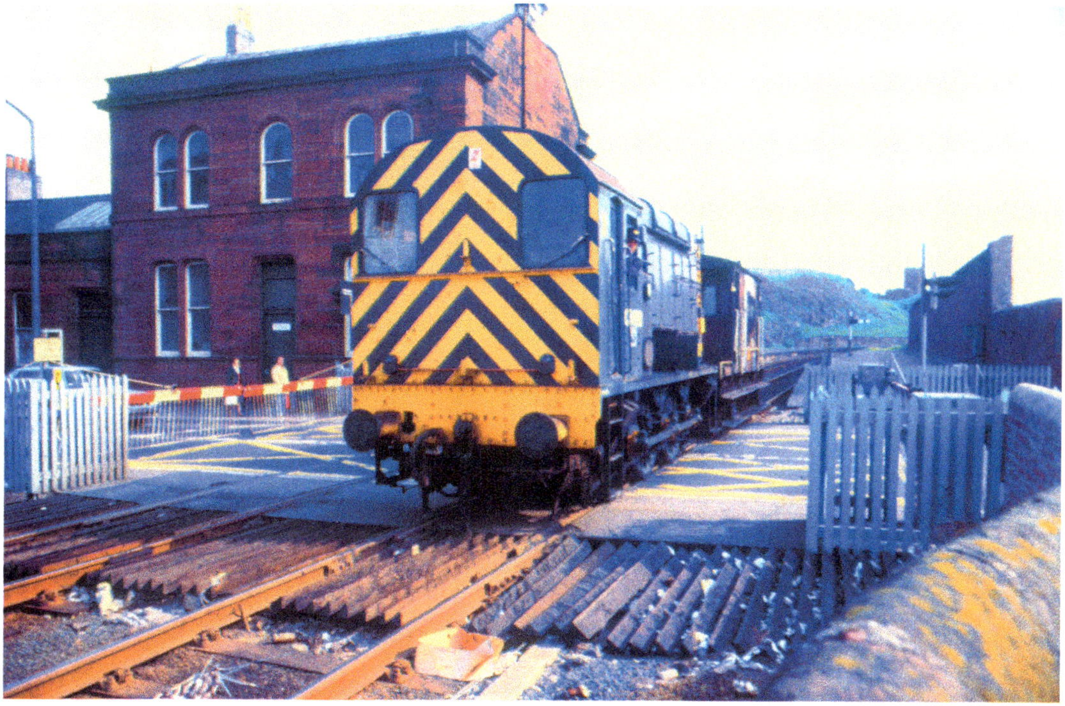

No. 08433 (AY) at Ardrossan town level crossing with a brake van heading home to Ardrossan Harbour Sidings. Taken September 1980.

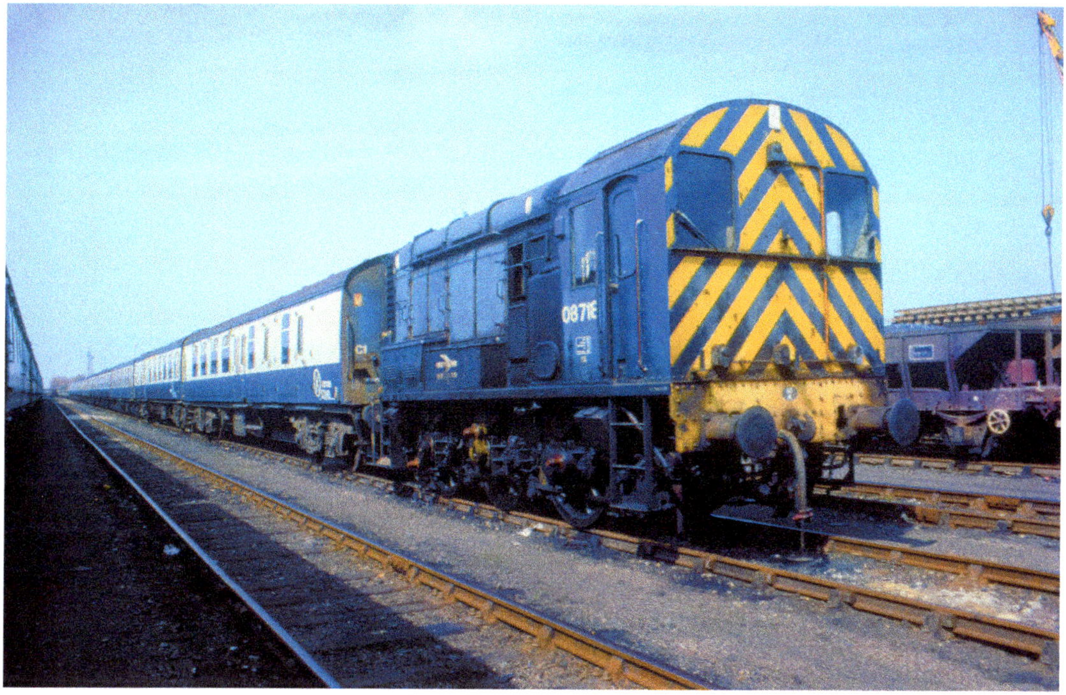

No. 08718 (HA) at Millerhill Yard with condemned Mk 1 rolling stock. Taken May 1983.

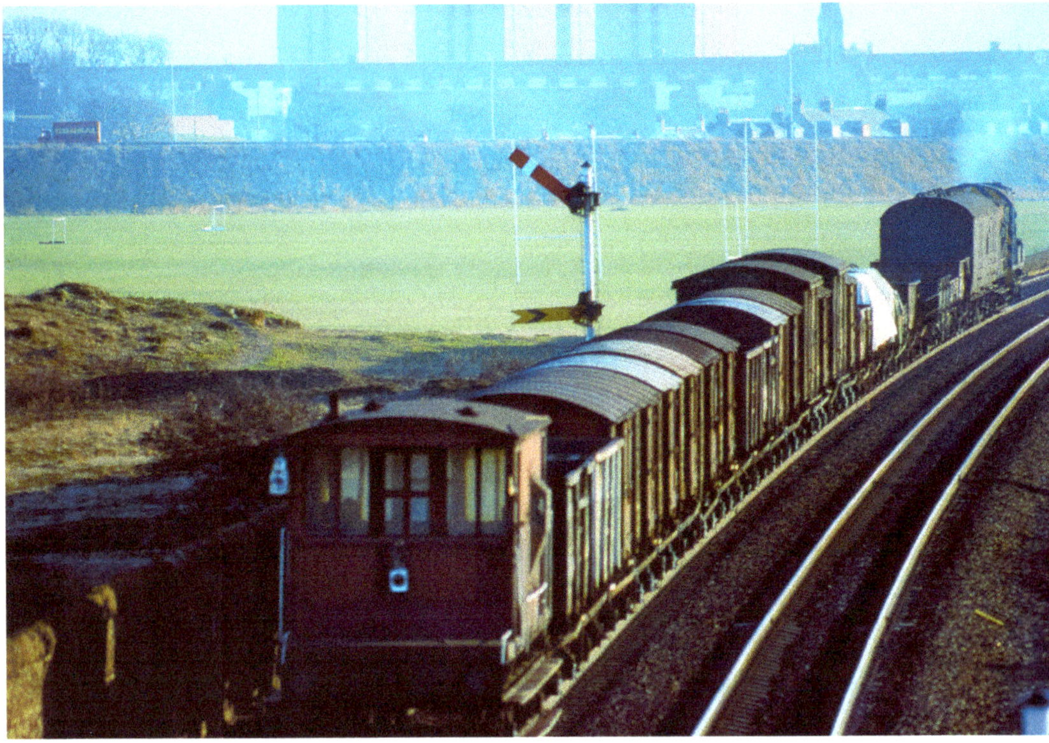

An unidentified Class 08 with local trip working at Bogside, north of Irvine. This loco would probably be heading to Falkland Yard via the Troon Avoiding Line. Taken February 1978.

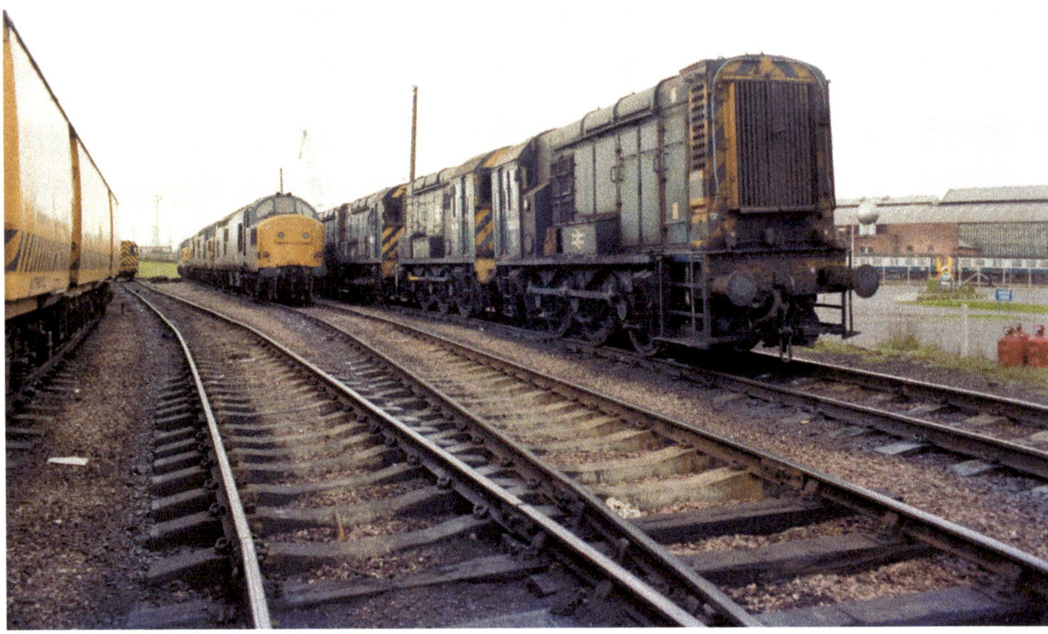

No. 08312 (ML) plus two other classmates and Class 37s stabled at Motherwell Depot. Taken April 1982.

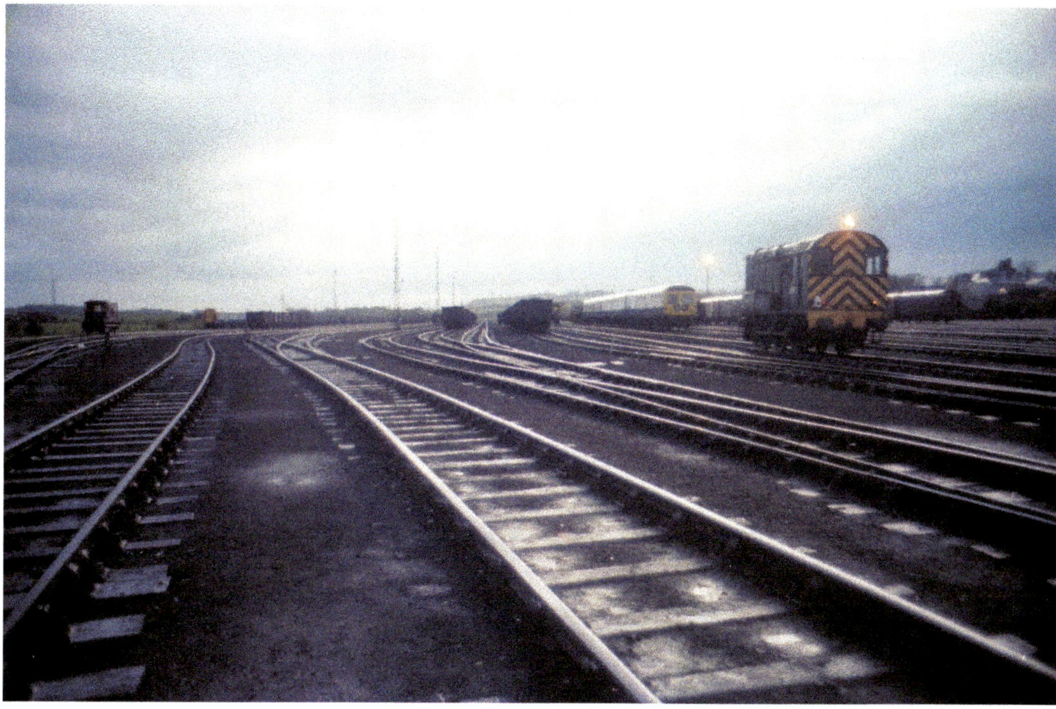

Class 08 rests at Thornton Yard between MGR shunting duties. Note also two three-car Class 126 DMUs – one for preservation and the other for the torch man. Taken May 1983.

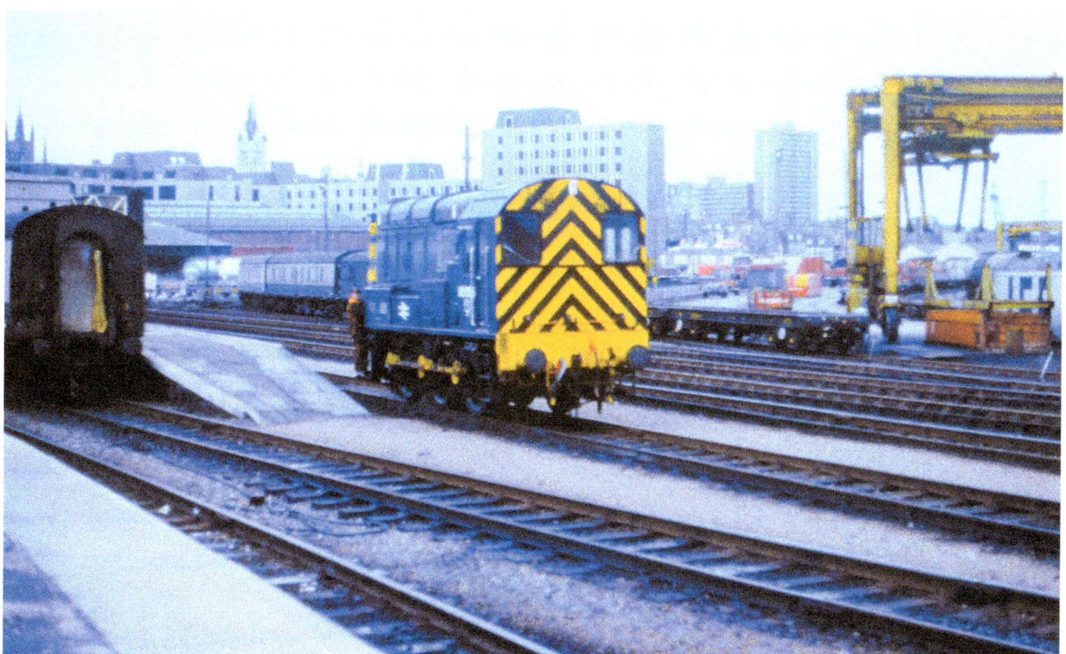

No. 08855 (AB) at Aberdeen station resting between shunting duties at the nearby Freightliner depot at Guild Street. Taken February 1984.

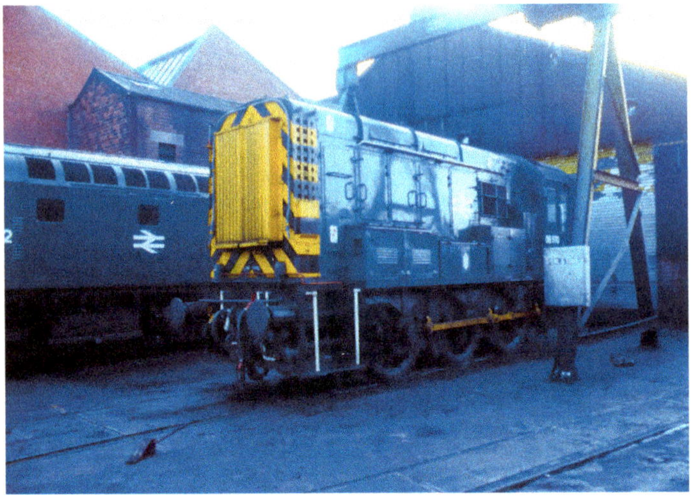

No. 08570 (HA) at Haymarket Depot. This loco was withdrawn from service in November 1992 and scrapped at Springburn Works by September 1993. Taken February 1984.

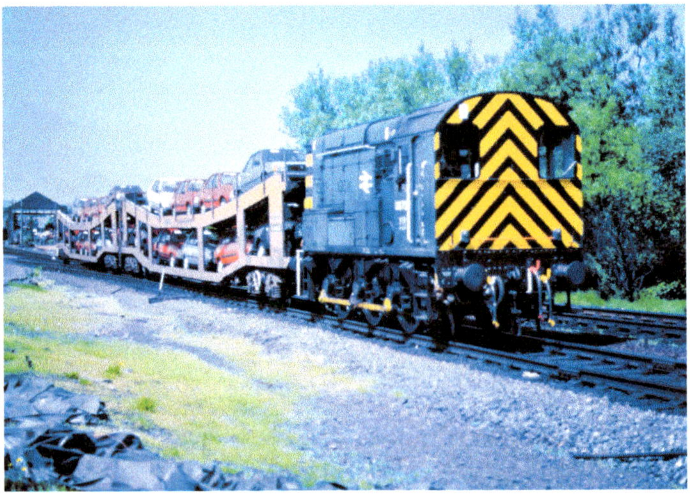

No. 08733 (AY) at Stranraer Town shunting car flats into the Stockton Haulage Sidings. These cars were later loaded onto a freight ship to Larne. Taken June 1984.

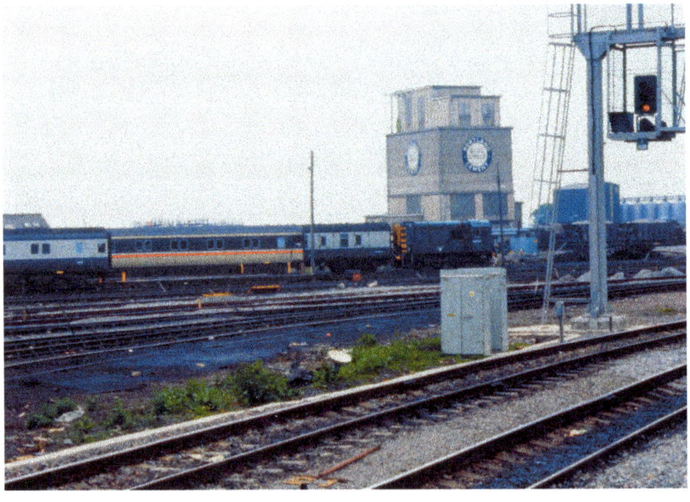

No. 08754 (IS) at Inverness Yard shunting passenger coaches. Taken July 1987

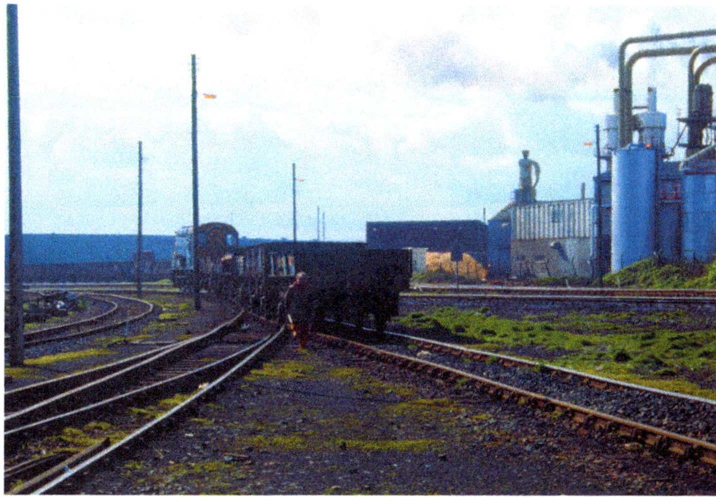

No. 08448 (AY) shunting engineers' wagons at Irvine East Yard. The driver is being assisted by a local chargeman. Taken April 1986.

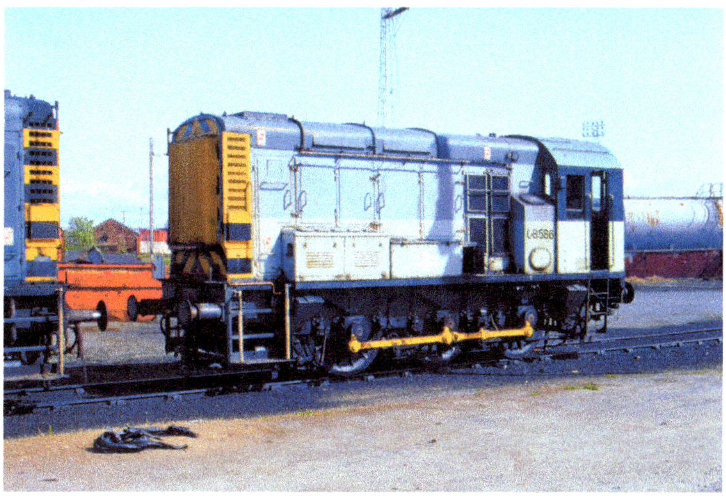

No. 08586 (AY) at Ayr Depot. This loco was withdrawn from service in October 1988 and was cut up on site at Ayr Depot by local contractors by May 2000. Taken May 1993.

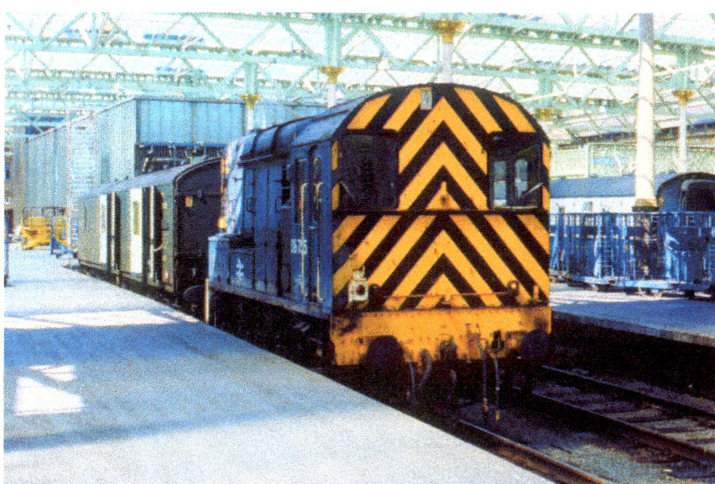

No. 08755 (HA) at Edinburgh Waverley East End shunting, probably with the previous night's newspaper vans. Taken May 1985.

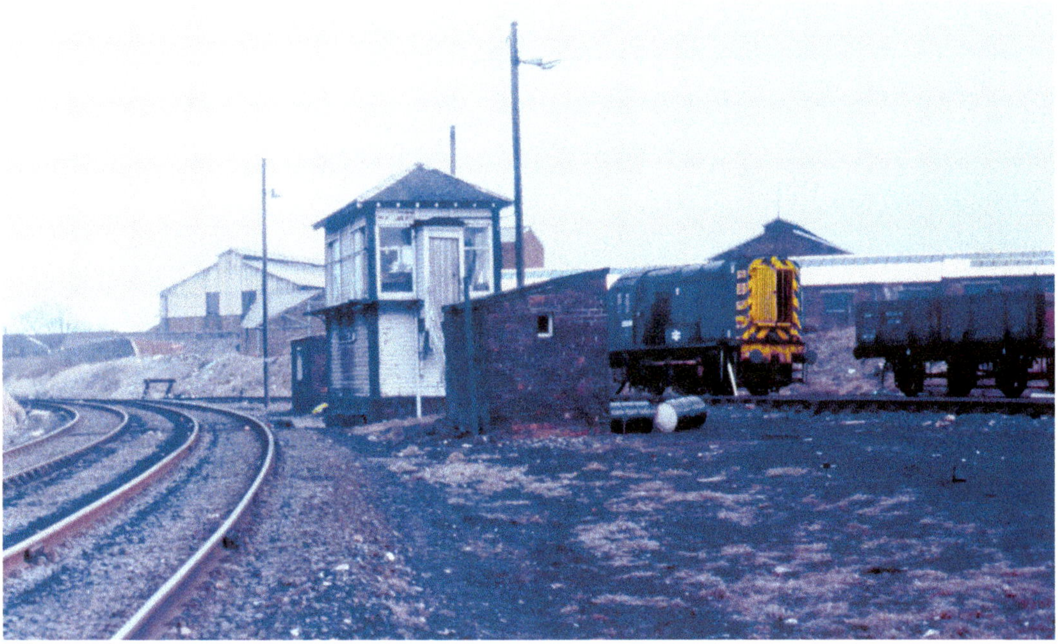

No. 08344 (AY) stabled at Ayr Depot. On the left was Blackhouse Junction signal box, which was on the then double track from Newton-on-Ayr Junction to Annbank Junction. This line was singled in May 1985 and the signal box was removed in August of the same year. Taken February 1985.

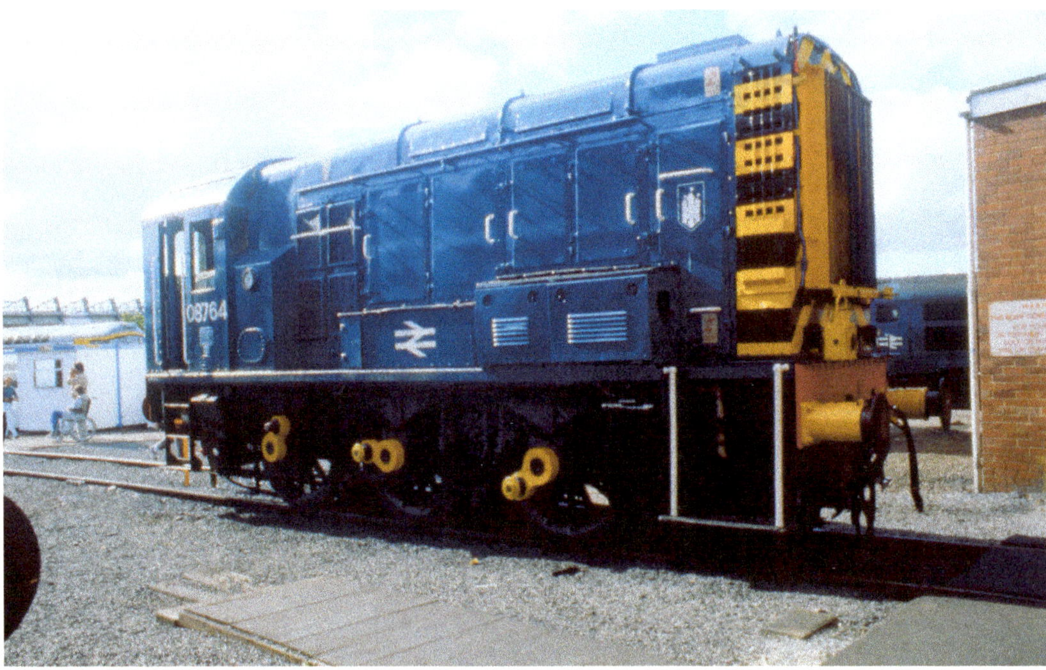

No. 08764 (HA) at Haymarket Depot, which was on display for a BR open day. Taken August 1985.

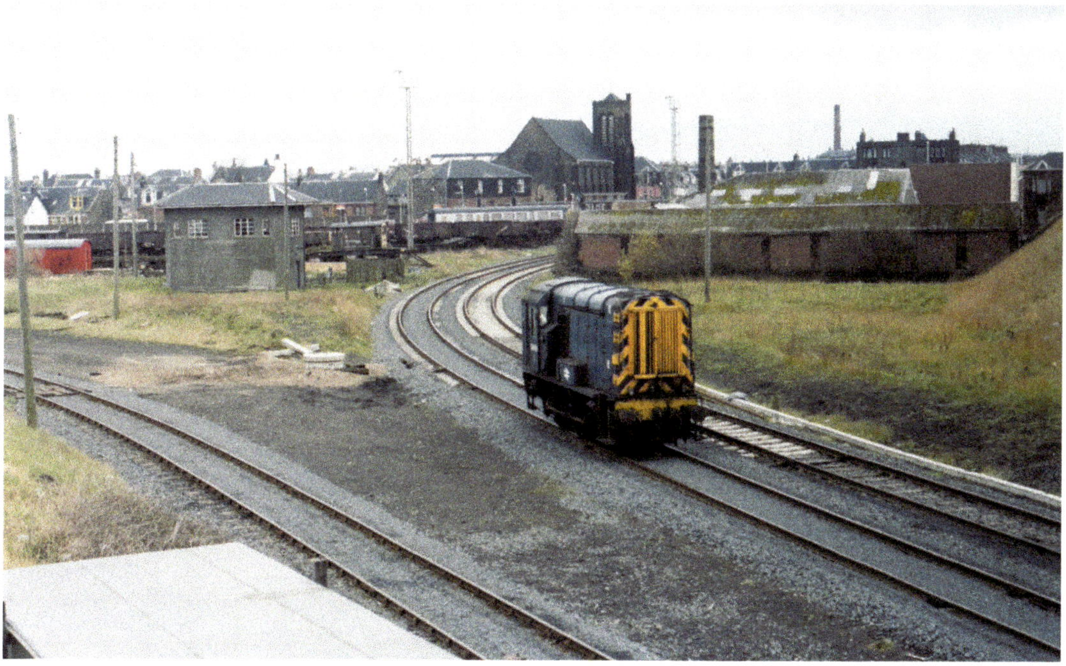

No. 08449 (AY) on the Ayr Harbour branch heading light diesel to Ayr Depot. Taken April 1985.

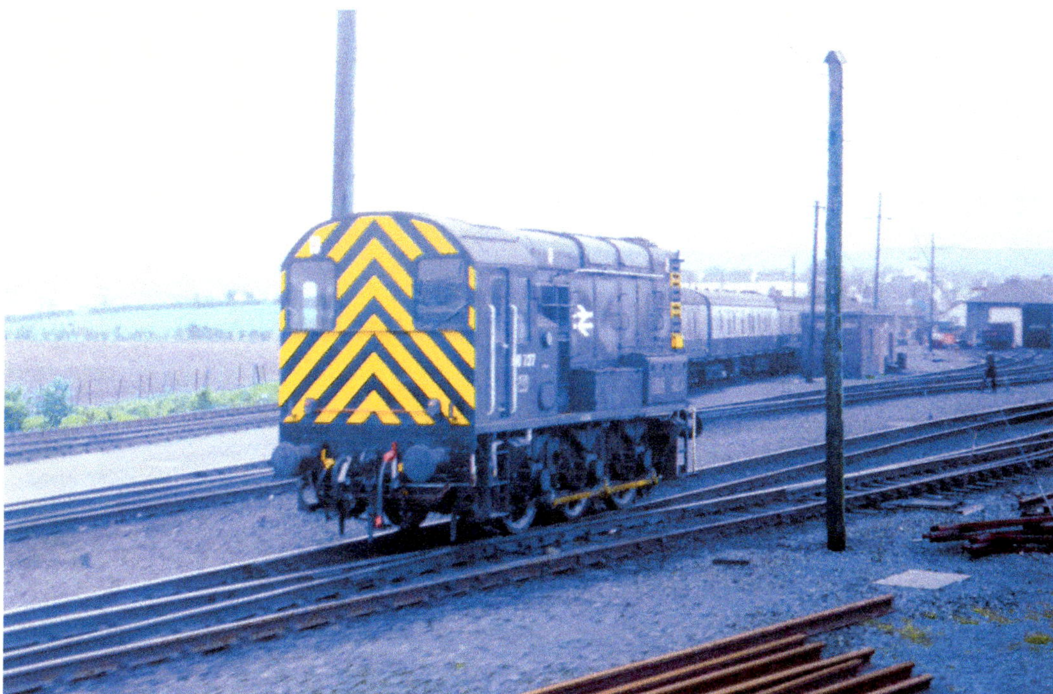

No. 08727 (AY) at Stranraer Town shunting Mk 3 Sleepers off the overnight Sleeper. This was amongst its duties at this time, as well as shunting wagons for the nearby Stockton Haulage Company. Taken May 1985.

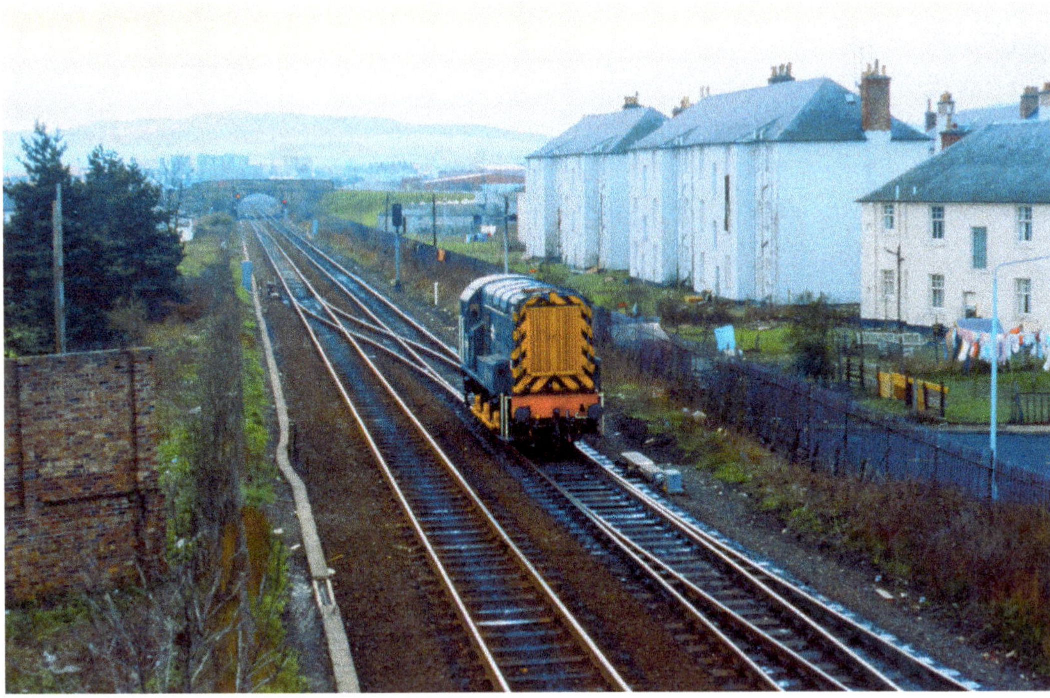

No. 08827 (TJ) approaching Perth Yard light diesel to collect wagons. Taken April 1985.

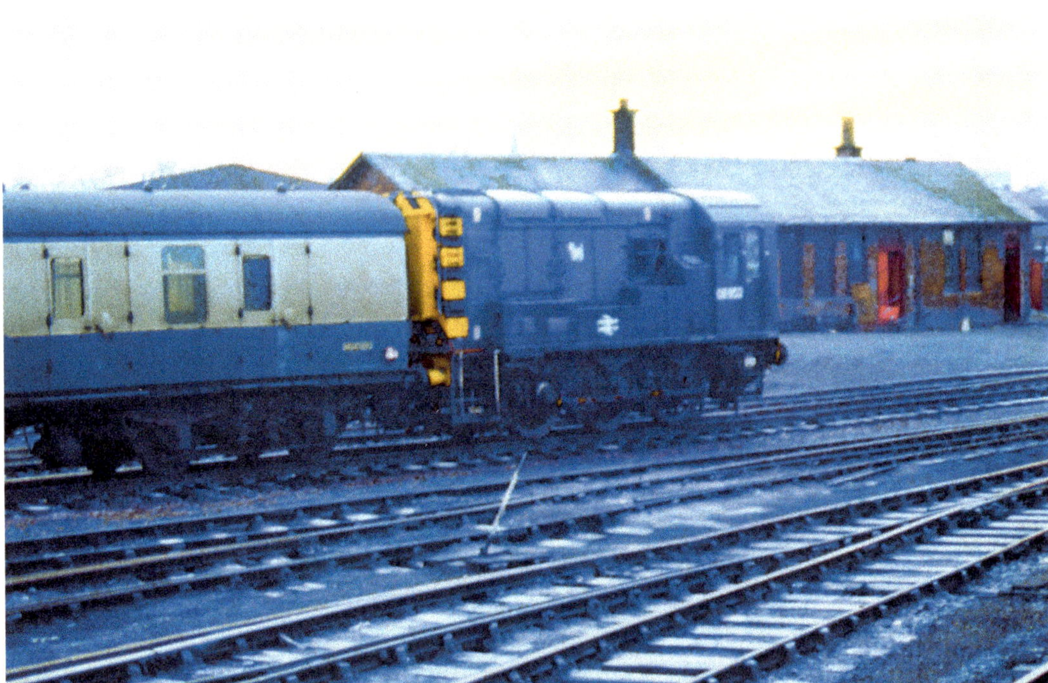

No. 08952 (ED) at Glasgow Salkeld Street Parcels Depot shunting parcel coaches. Taken February 1986.

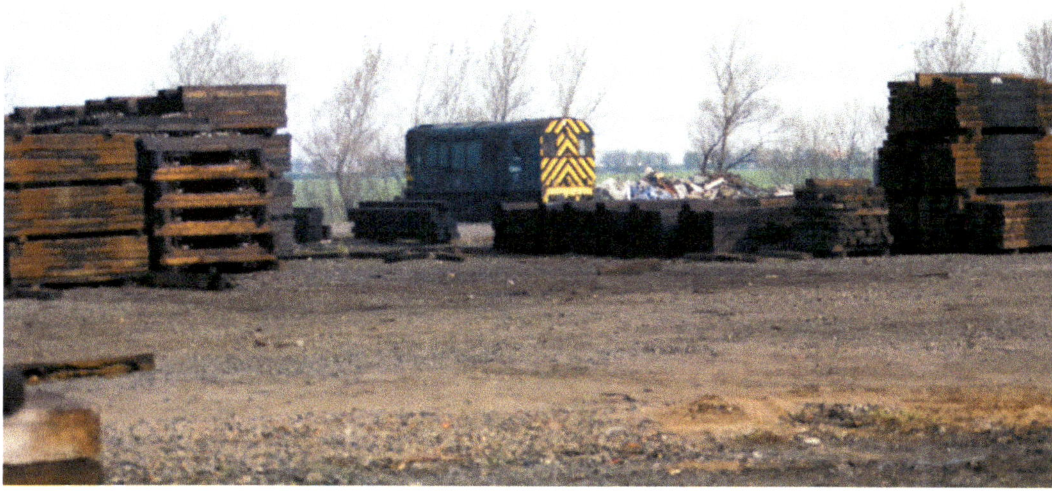

No. 08714 (HA) at Millerhill Depot fuelling point. Taken May 1983.

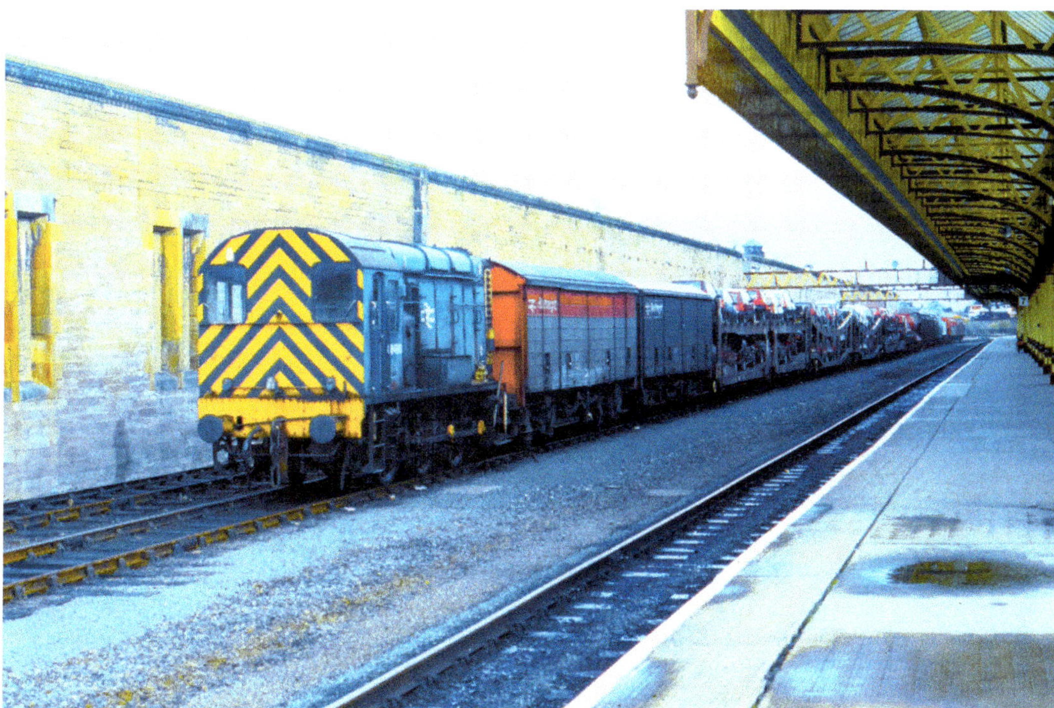

No. 08496 (TJ) at Perth station with a Speedlink demonstration train. This was in connection with another BR open day at Perth Yard. Taken April 1985.

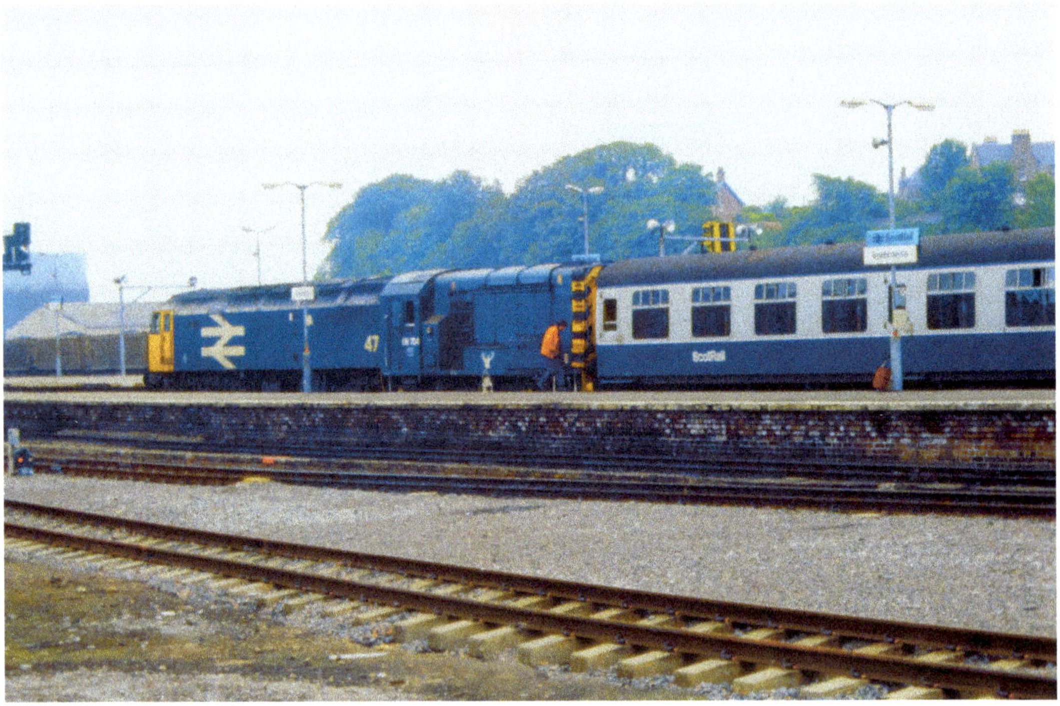

No. 08568 (IS) at Inverness station shunting Mk 2 coaches for cleaning at the nearby depot. Taken August 1984.

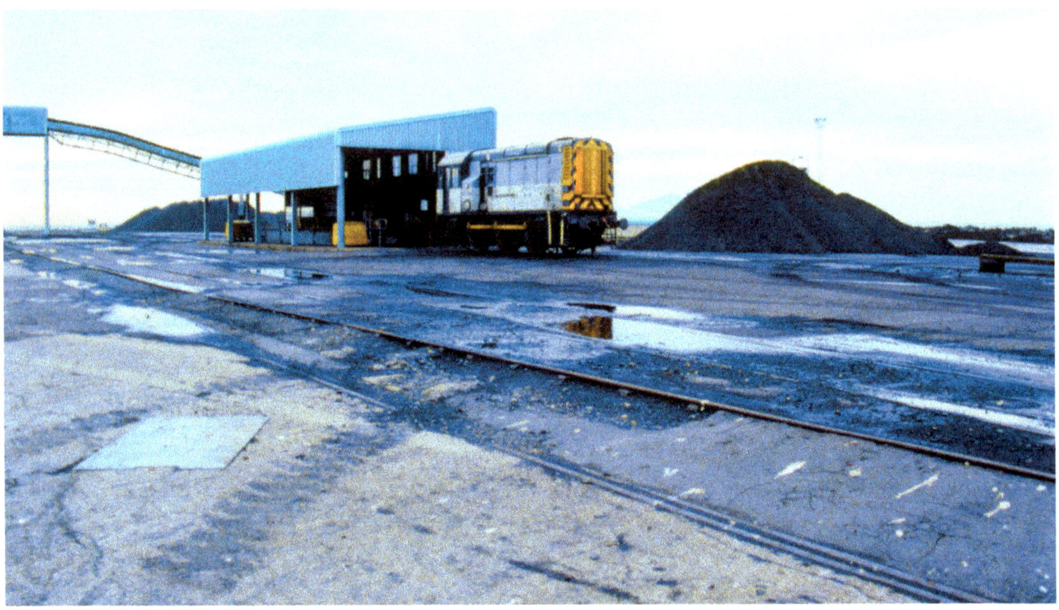

No. 08506 (AY) at the Ayr Harbour Coaling Plant. This is where coal from the Ayrshire mines was shipped overseas mainly to Northern Ireland. This location was originally served by the Ayr Waggonway rail system, which ran east from here to serve mines east of the town. Taken March 1992.

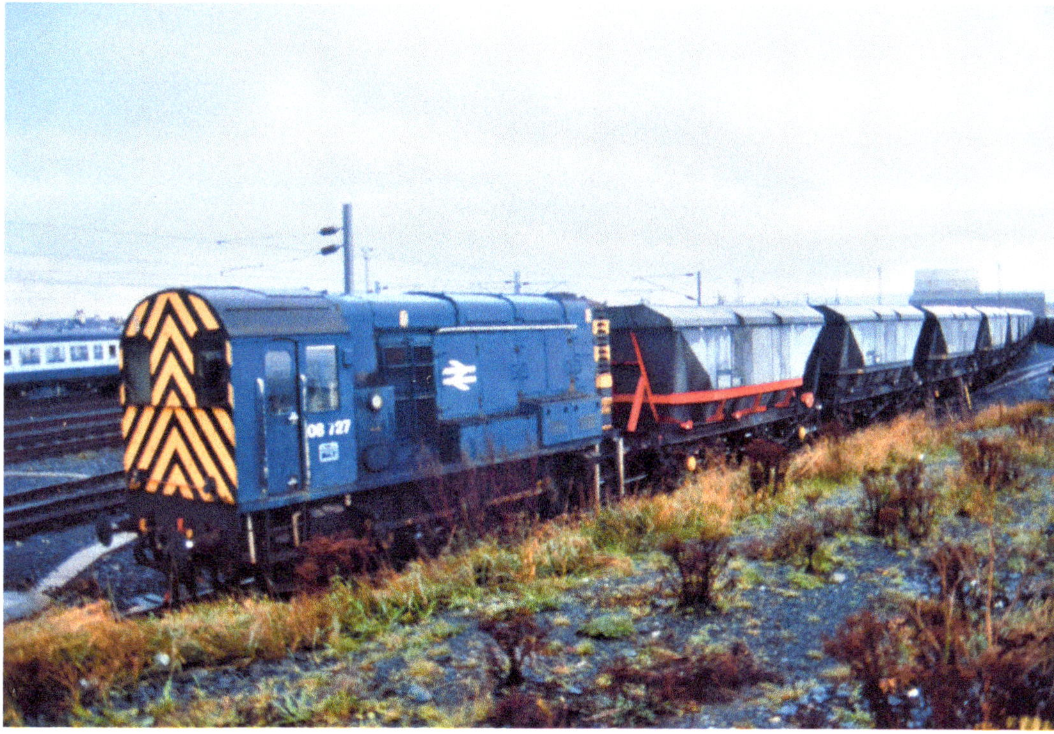

No. 08727 (AY) at Falkland Yard shunting MGR wagons. Taken August 1987.

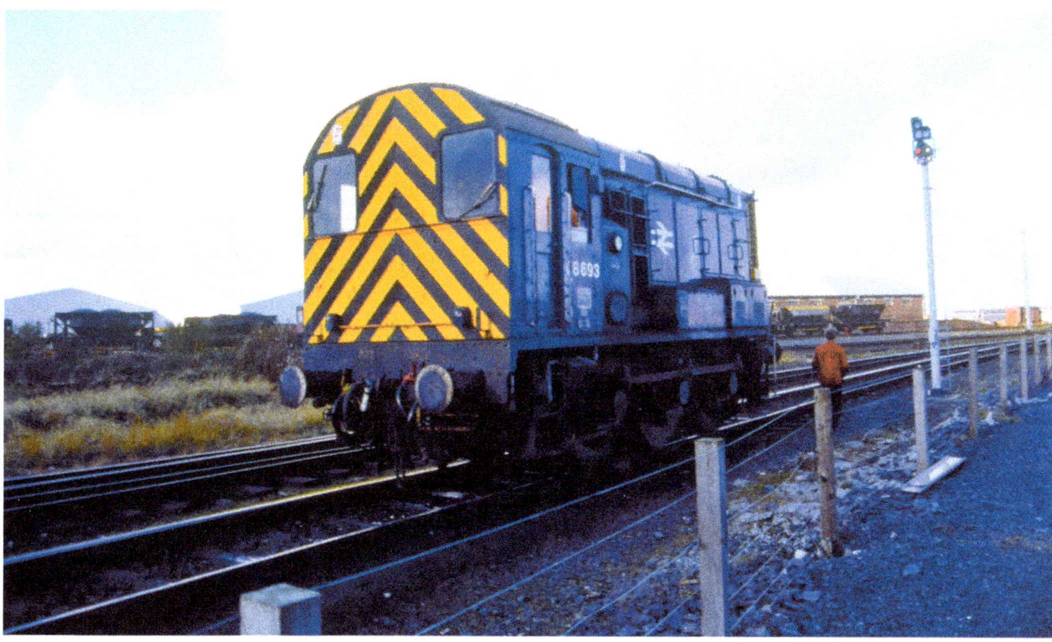

No. 08693 (AY) at Irvine East Yard. The loco was heading from the East Yard onto the main Ayr to Glasgow Central line towards Irvine station. Behind the loco, this line ran to Kilmarnock until the Beeching axe closed it in 1965. Taken August 1987.

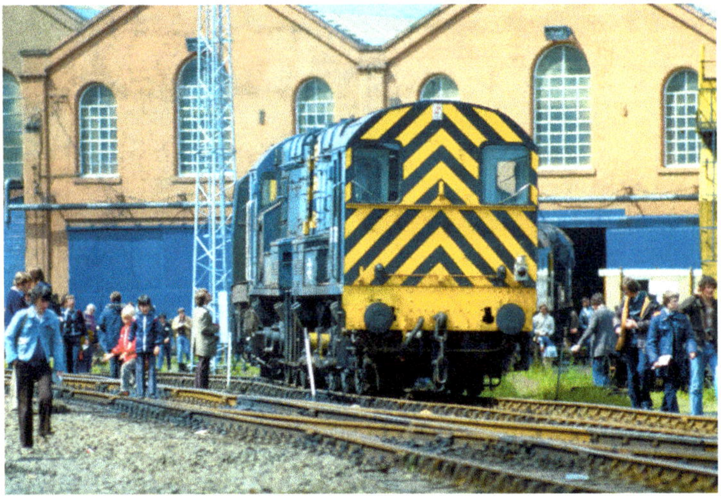

Class 08s at Springburn Works, Glasgow, on display for part of a BR works open day. Taken June 1981.

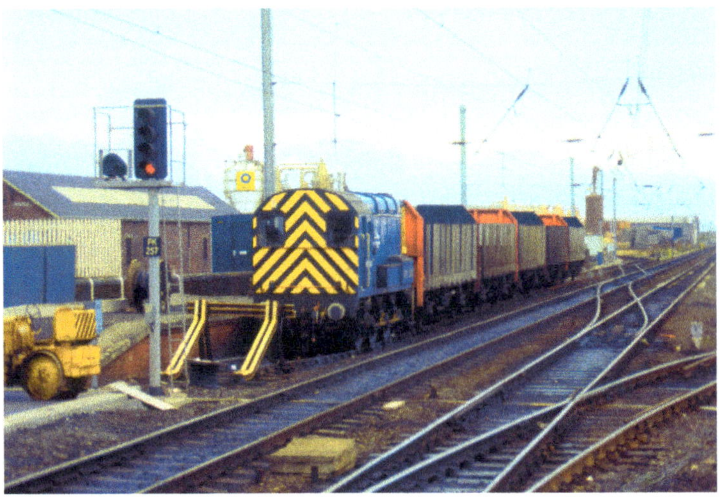

No. 08693 (AY) at Irvine station shunting timber wagons from the nearby Caperboard factory in the east yard. This factory, which processed timber, closed in 1996. Taken November 1986.

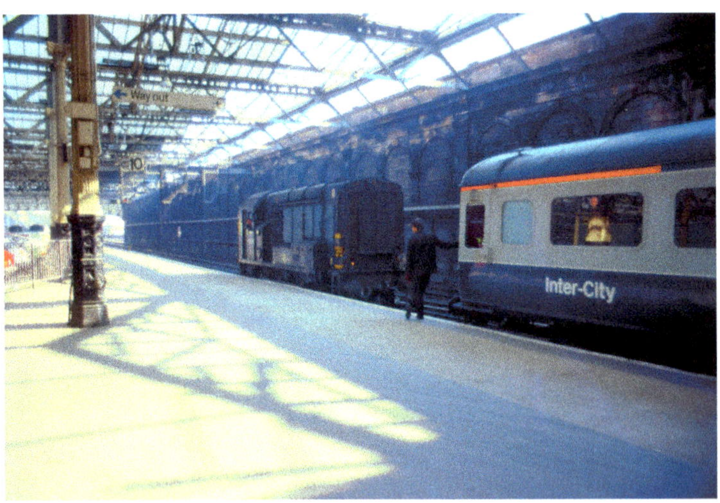

No. 08881 (HA) at Edinburgh Waverley about to attach to Mk 2 coaching stock off a CrossCountry InterCity service and take the stock to Craigentinny Depot for cleaning. Taken March 1982.

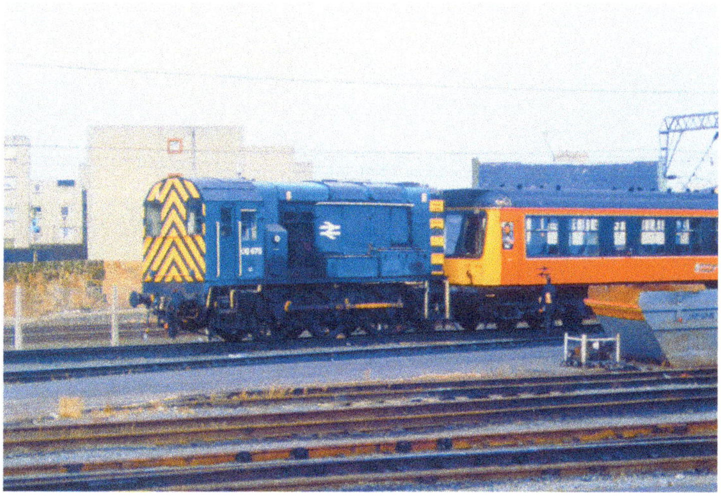

No. 08675 (AY) at Ayr Depot shunting a Class 107 power car. Taken August 1987.

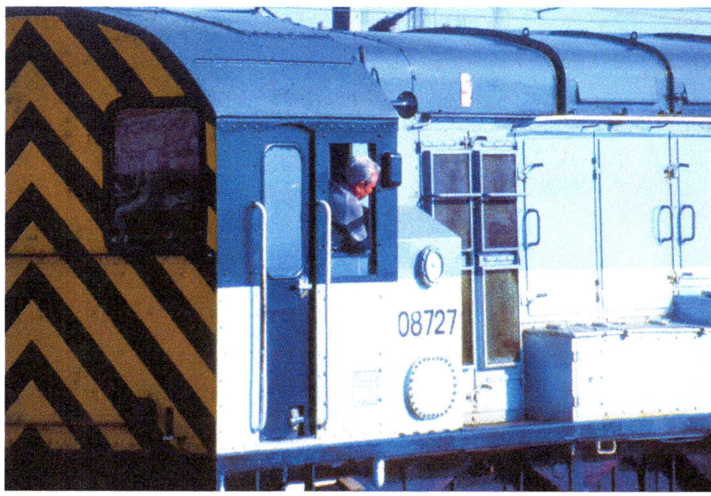

A close up of driver Jock McTavish of Ayr Depot at the controls of No. 08727 (AY) at Falkland Yard. Taken May 1990.

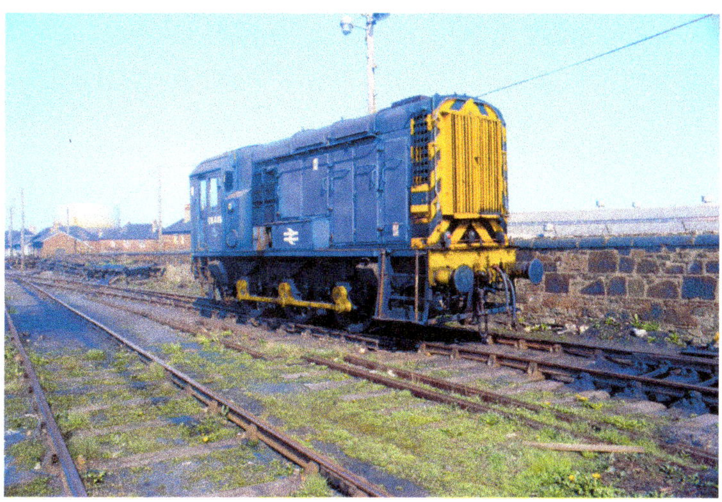

No. 08675 (AY) stabled at Ayr Harbour between duties. This loco was transferred to Motherwell Depot in July 1995. Taken April 1987.

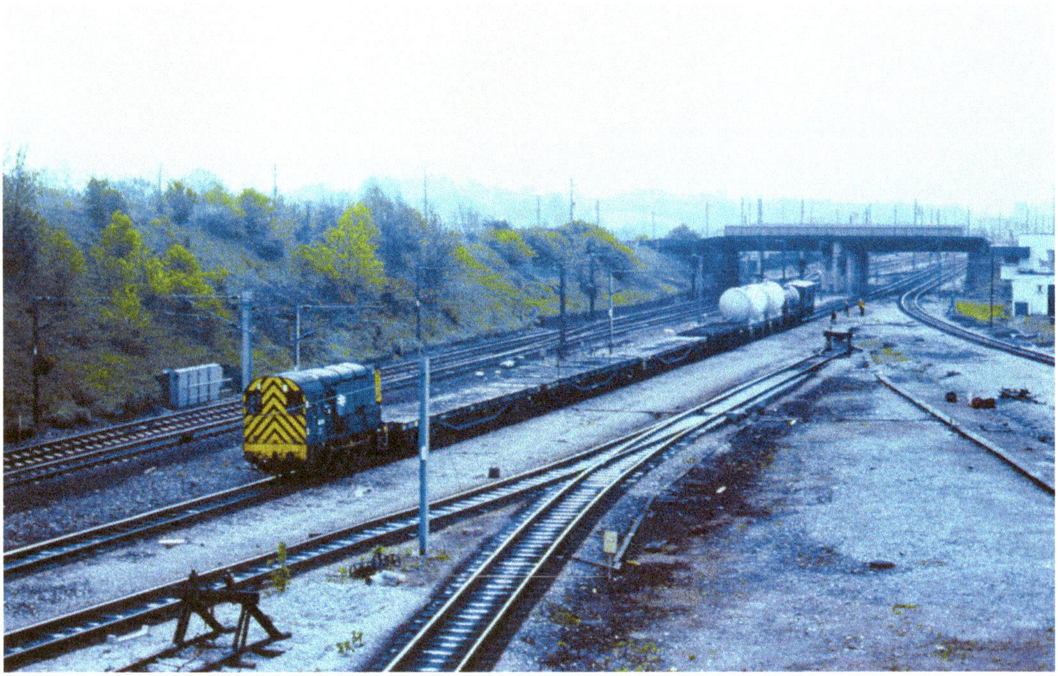

No. 08911 (KD) at Carlisle Kingmoor Yard with a local transfer freight arriving in at the south end. Many freight services were remarshalled here prior to working onwards to Scotland. Taken May 1983.

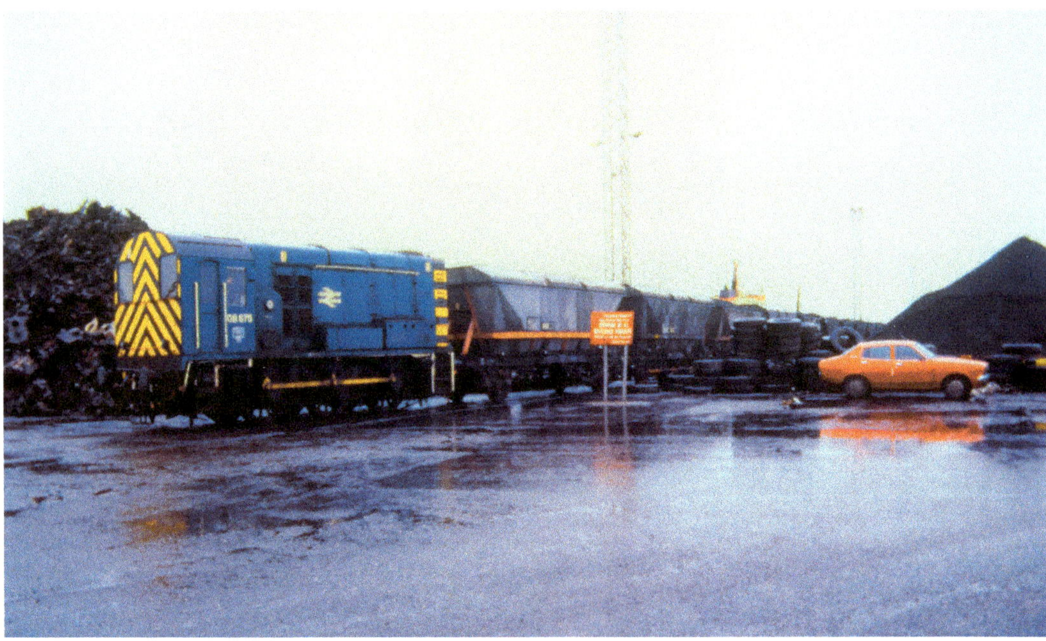

No. 08675 (AY) at Ayr Harbour shunting MGR wagons. This area of the harbour was also used for the recycling of scrap metal. The Ayr-based loco shunters were often involved in moving scrap wagons as well as MGR wagons. Taken April 1987.

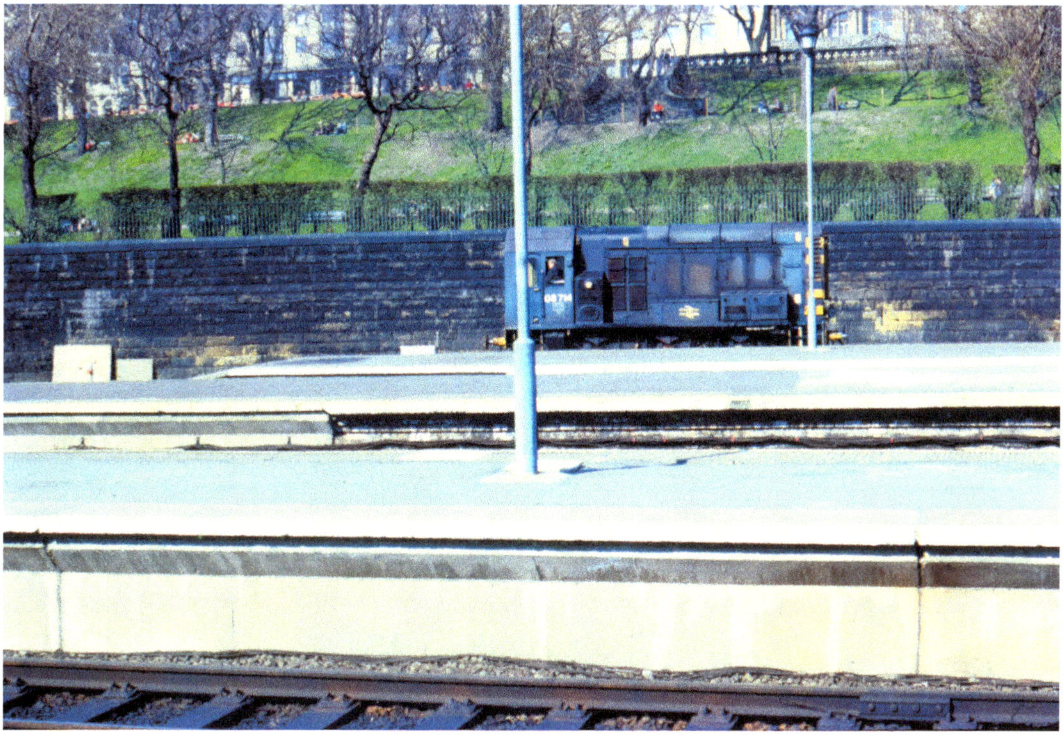

No. 08714 (HA) at Edinburgh en route to Haymarket Depot. Taken March 1982.

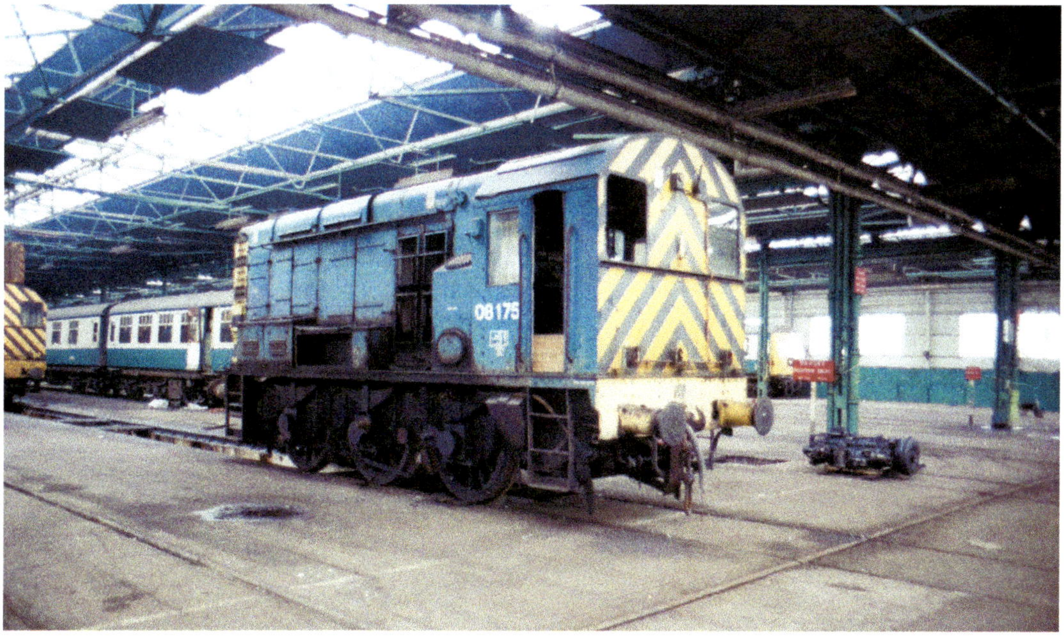

No. 08175 (TJ), withdrawn from traffic, stabled inside Dunfermline Townhill Depot. This loco was withdrawn from traffic in December 1984 and scrapped at JR Adams of Glasgow by April 1986. Taken April 1983.

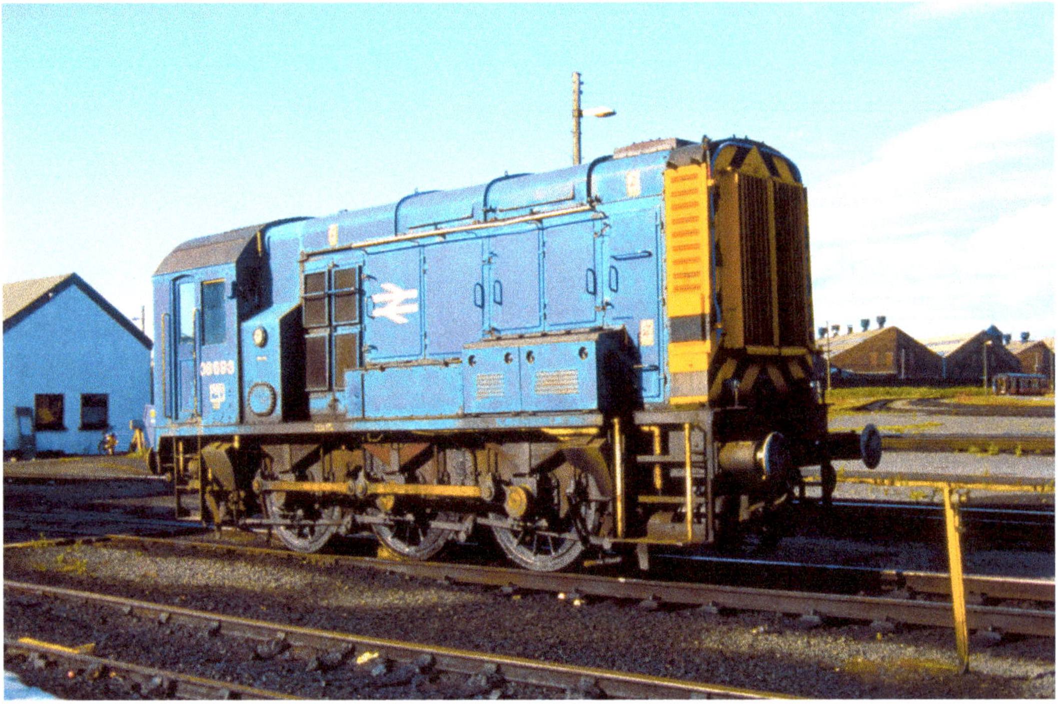

No. 08693 (AY) at Ayr Depot. This loco was transferred to Motherwell Depot in June 1990 and was eventually scrapped at Wigan CRDC by May 2000. Taken September 1987.

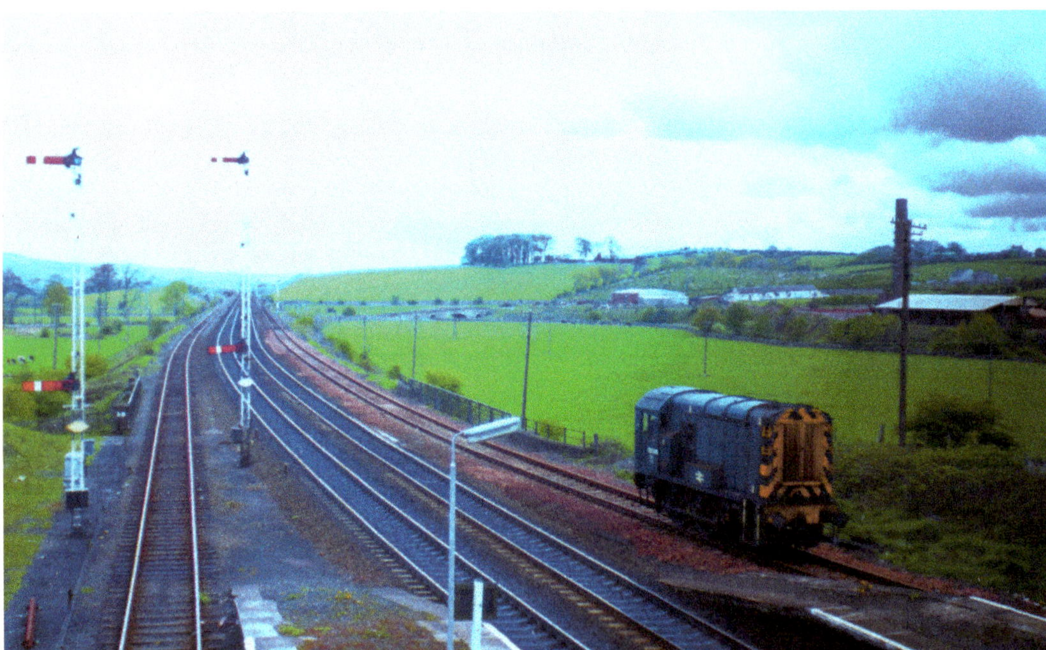

No. 08345 (AY) on the Up Slow line approaching Dalry station. This loco was originally numbered D3415 and was renumbered to 08345 in September 1974. Taken May 1978.

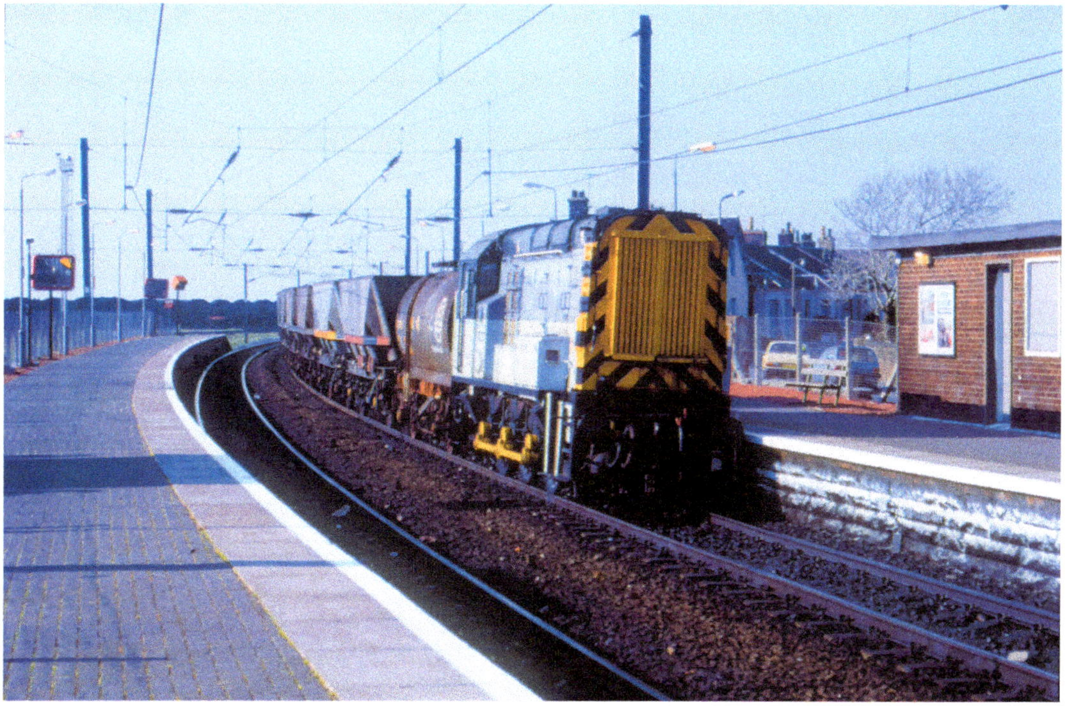

No. 08727 at Newton-on-Ayr station with a local transfer trip working from Falkland Yard to Ayr Depot. The wagons would be taken to the depot for repairs. Taken May 1990.

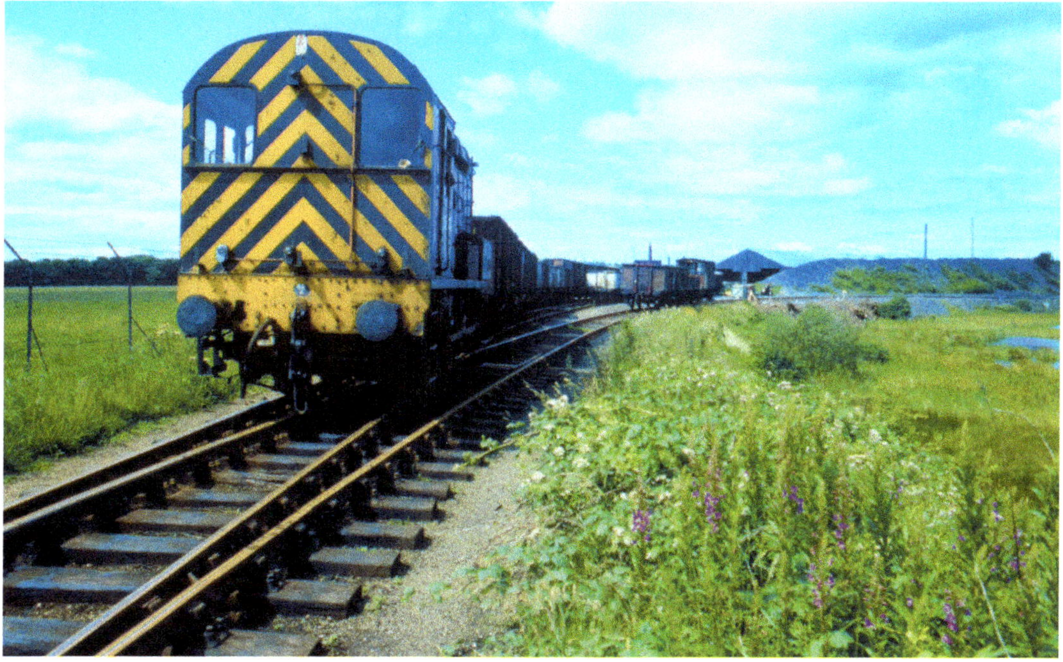

No. 08731 (AY) at ICI Snodgrass, between Irvine and Kilwinning, shunting gunpowder wagons. Taken September 1980.

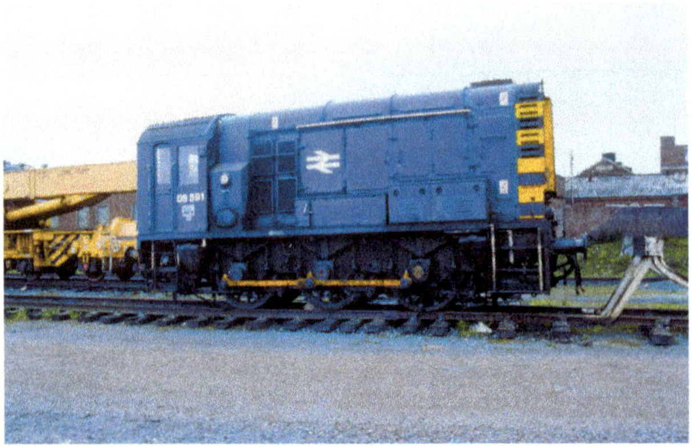

No. 08591 (AY) at Ayr Depot. This loco was withdrawn from traffic in February 1993 and finally scrapped by November of the same year. Taken June 1991.

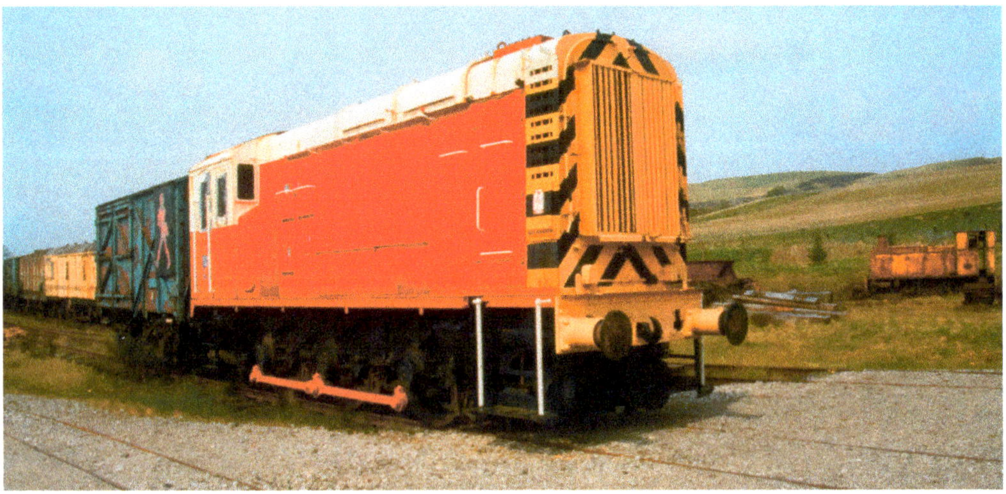

Similar in appearance to a Class 08, this loco is in fact a Class 11. It was originally numbered 12052 and worked at various BR depots, mainly in the north-east of England. It was eventually sold to the Ayrshire Railway Preservation Group in October 1988, but was later sold on to the Brechin Caledonian Railway in July 2002. It is still currently undergoing restoration. The loco was photographed at the Ayrshire Railway Preservation Group's former residence of Minnivey, near Dalmellington, in East Ayrshire. Taken June 2002.

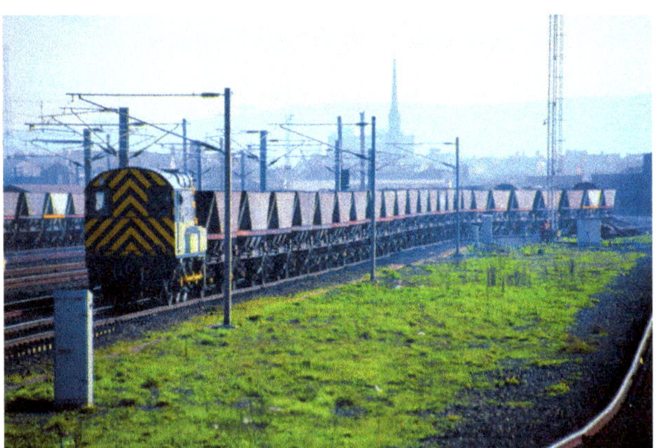

No. 08807 (ML) at Falkland Yard with MGR wagons. Taken November 2001.

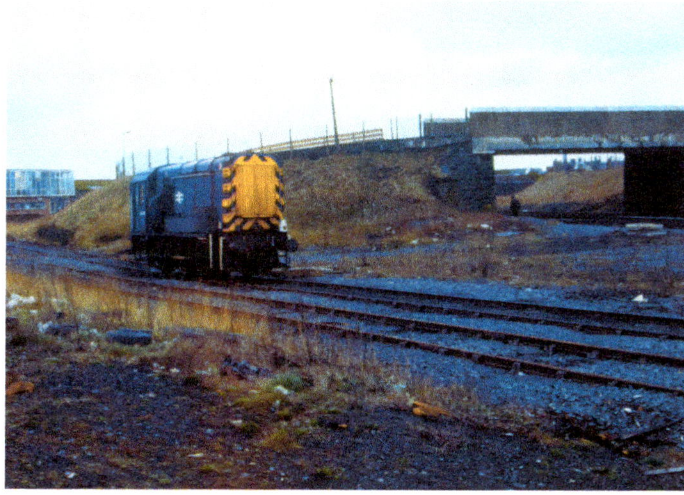

No. 08449 (AY) at Ayr Harbour stabled between shunting duties. These tracks led to the Scottish Agricultural Industries Factory but were removed in the late 1970s. This loco was withdrawn from traffic in May 1987 and eventually scrapped at European Metals Recycling by December 2001. Taken May 1984.

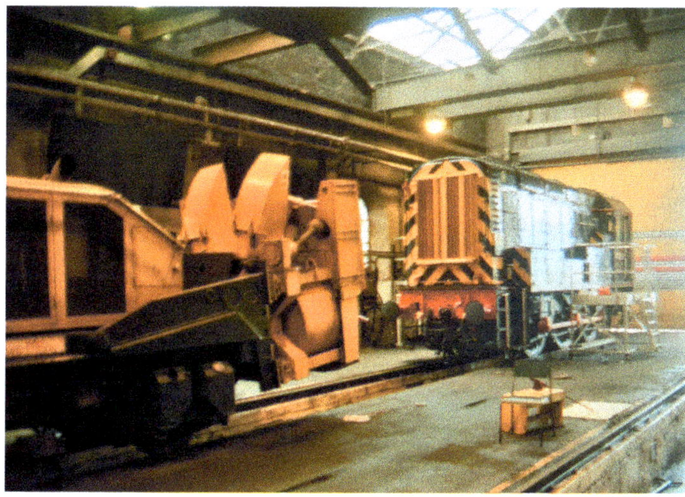

A Class 08 inside Inverness Depot stabled alongside one of the resident snow plough machines. Taken February 2000.

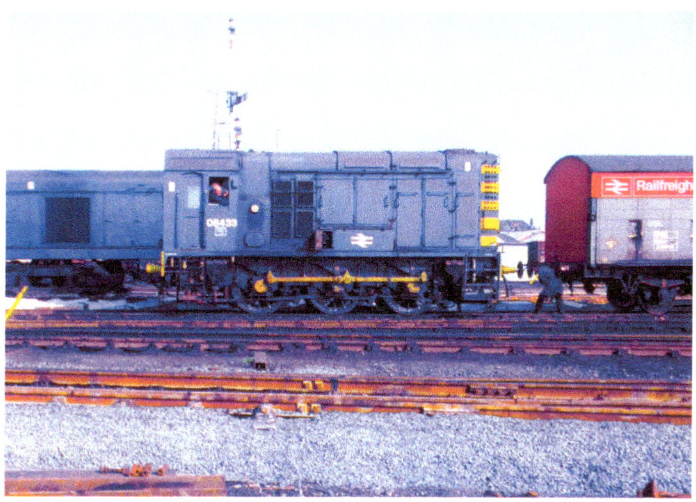

No. 08433 (AY) coupling up to a rail freight van at Falkland Yard. Freight vans are rarely seen on the railway nowadays. Taken April 1984.

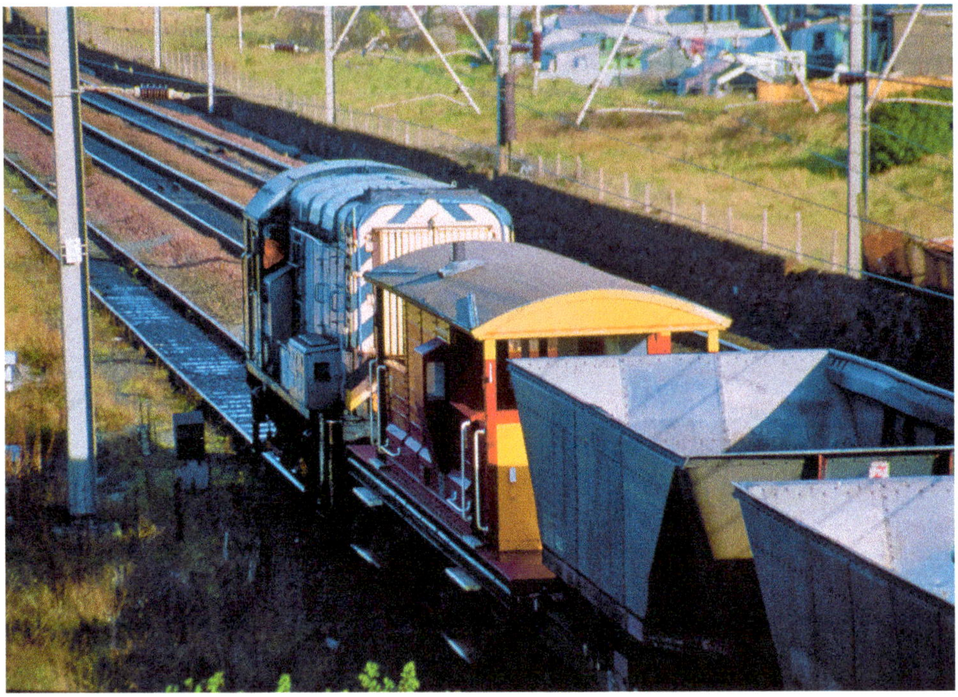

No. 08727 (AY) with a special brake van at the north end of Falkland Yard. These double-bogied brake vans were used on the Broomhill branch south of Ayr due to the trains having to be propelled off the Chalmerston branch. This small open cast mine only operated from 1997 to 2001. Taken January 1998.

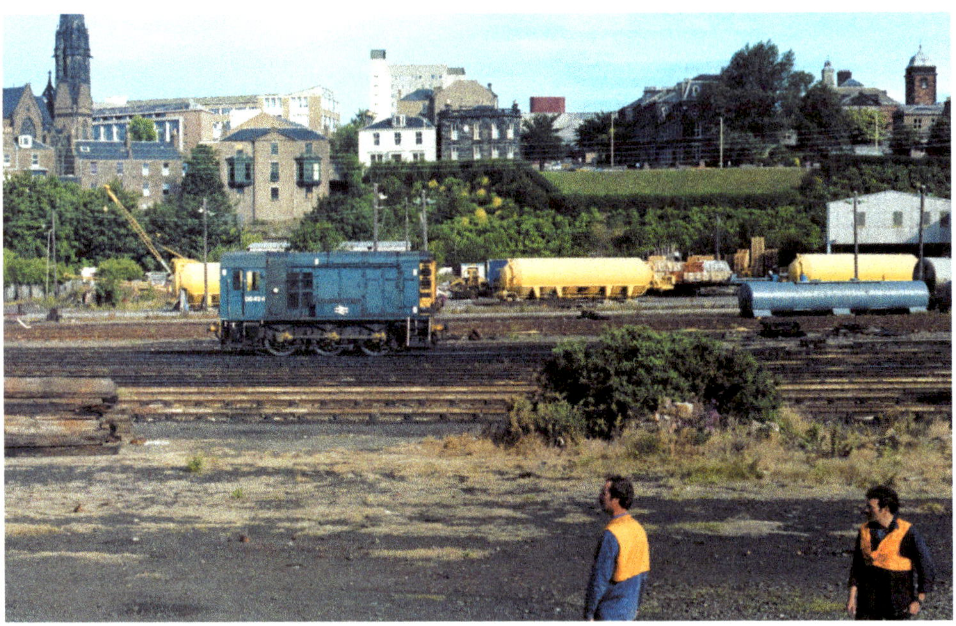

No. 08424 (DE) shunting in Dundee Yard. It was withdrawn from traffic in June 1985 and scrapped at London Road Goods Depot Yard in Glasgow by April 1986. Taken August 1981.

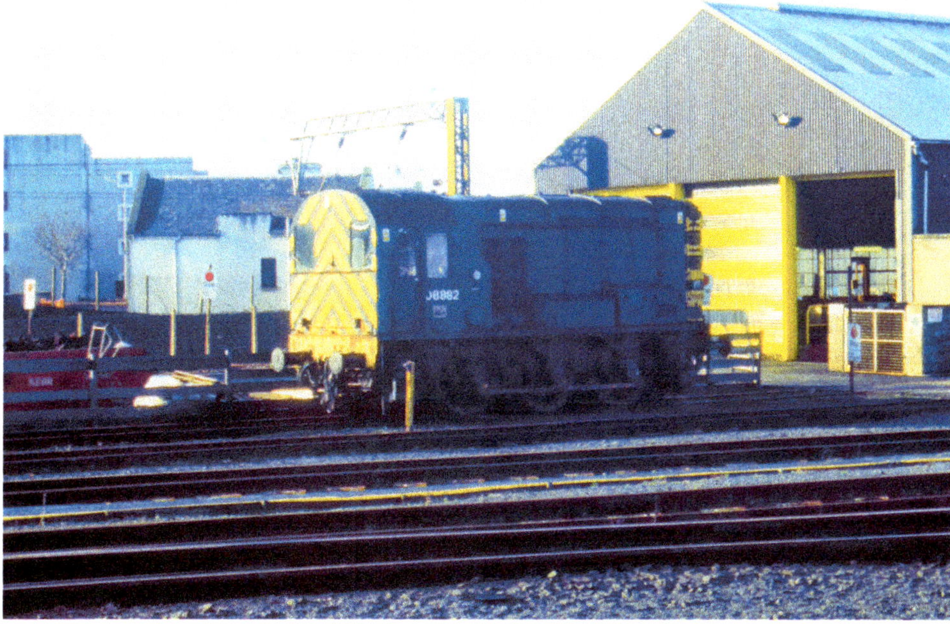

No. 08882 (ML) at Ayr Depot. This loco was withdrawn from traffic after sustaining fire damage at Motherwell Depot in August 2001 and was scrapped by January 2002. Taken December 1999.

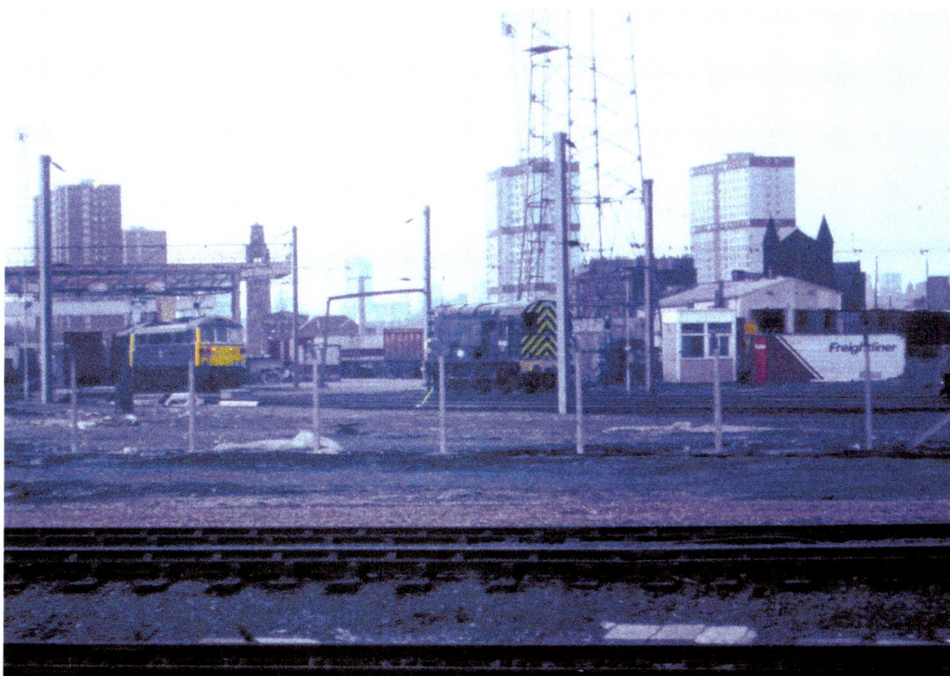

No. 08620 (ED) at Gushetfaulds Freightliner Depot, Glasgow, with two Class 86s for company. Gushetfaulds was closed in 1989 and all Freightliner operations concentrated at Coatbridge. Taken April 1983.

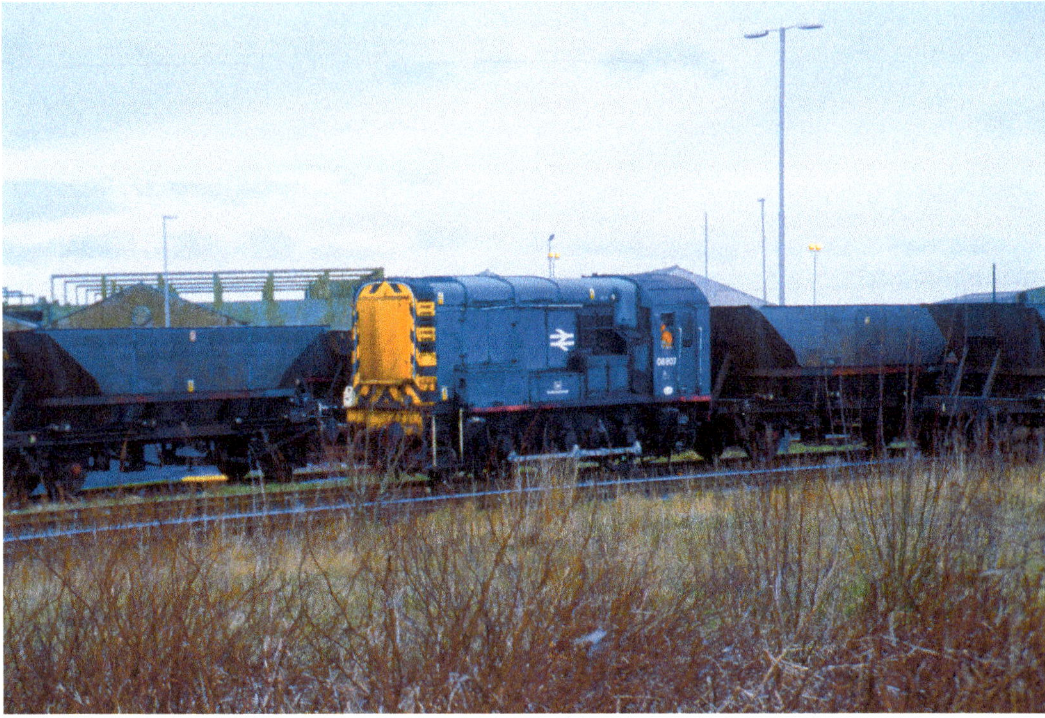

No. 08807 (AY) shunts a rake of empty MGR wagons at Falkland Yard. Taken May 1990.

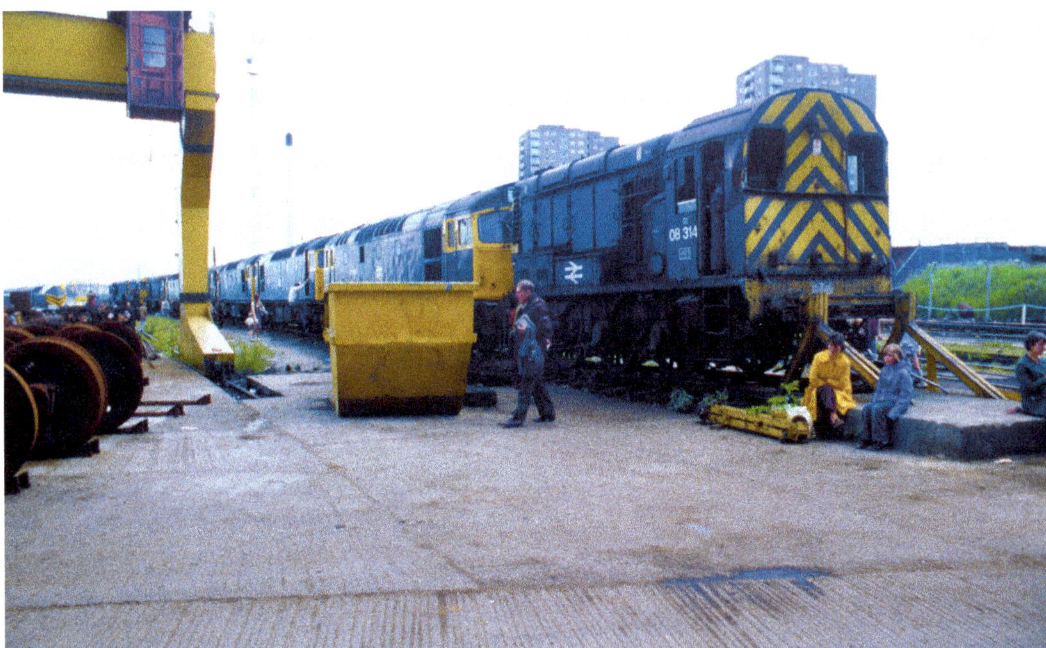

No. 08314 (HQ), already withdrawn from traffic, heads a line up of Class 26 and 27s at Springburn Works, Glasgow. Taken August 1981.

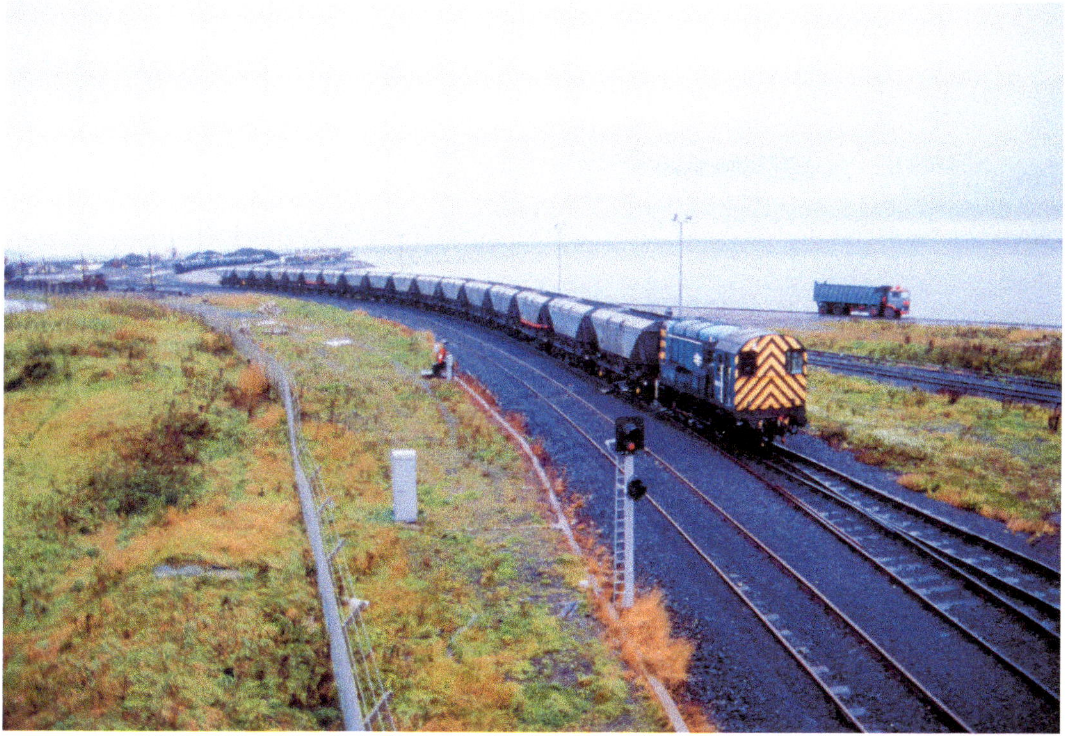

Above: No. 08675 (AY) at Ayr Harbour shunting loaded MGR wagons. Taken August 1987.

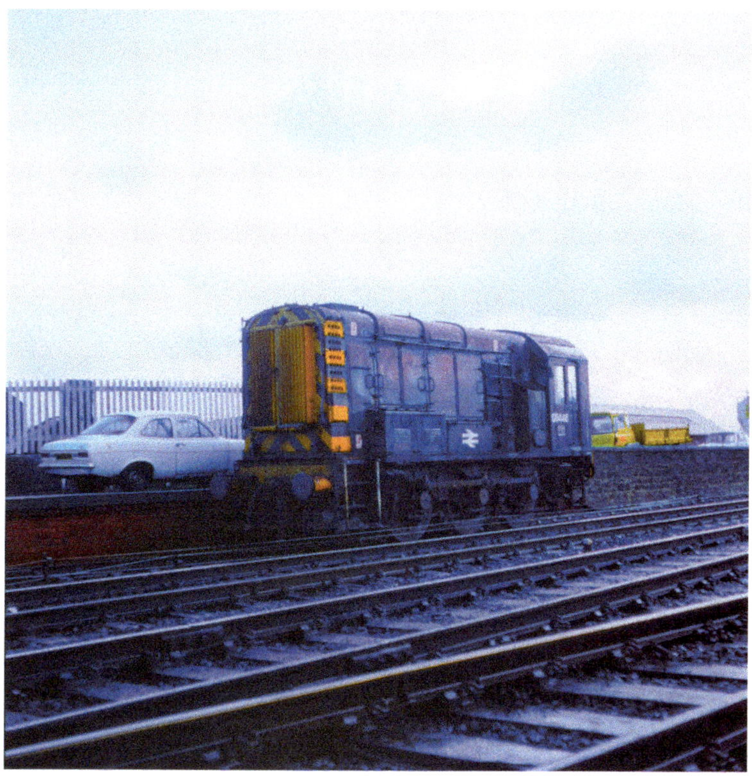

Right: No. 08446 (AY) stabled at Irvine station. This loco was transferred to Motherwell Depot in May 1984 and scrapped at J. McWilliams of Shettleston, Glasgow, by March 1986. Taken December 1981.

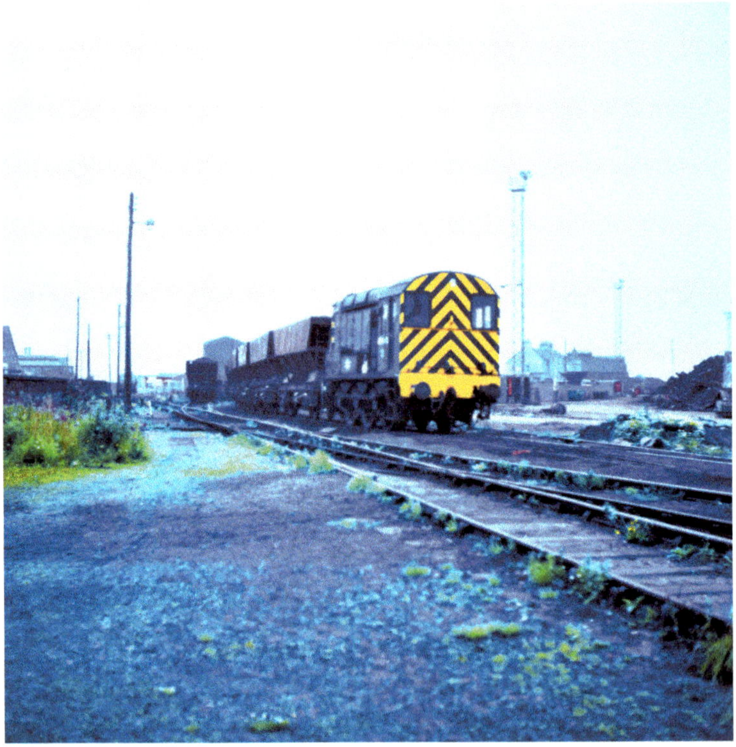

No. 08430 (AY) at Ayr Harbour with HTA wagons recently arrived off the Chalmerston branch. Taken August 1981.

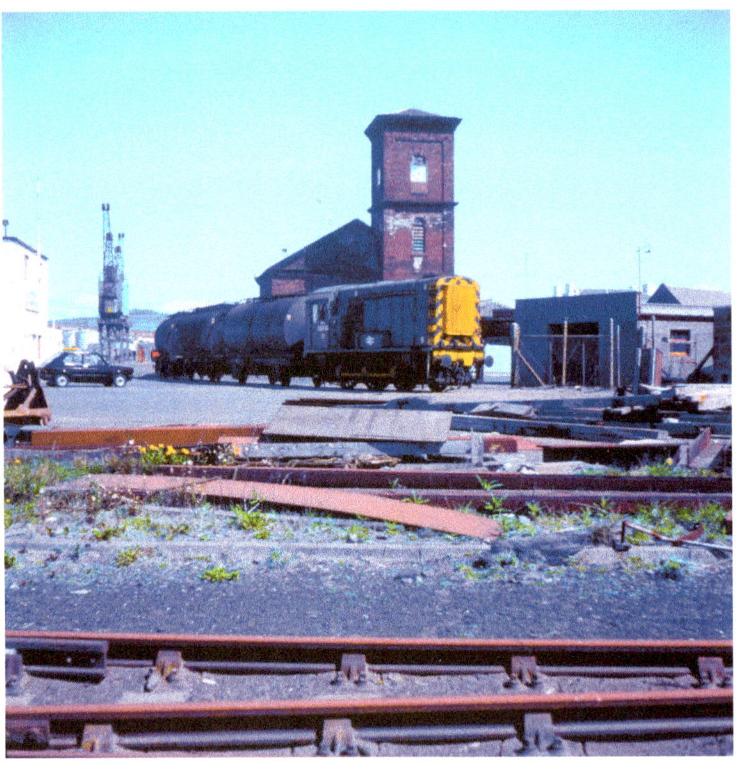

No. 08430 (AY) with bitumen tanks at Ardrossan Harbour from the nearby Shell Sidings. These tanks would be shunted and prepared for an onward service to Inverness Culloden Moor. This traffic ceased in the mid-1980s. Taken April 1984.

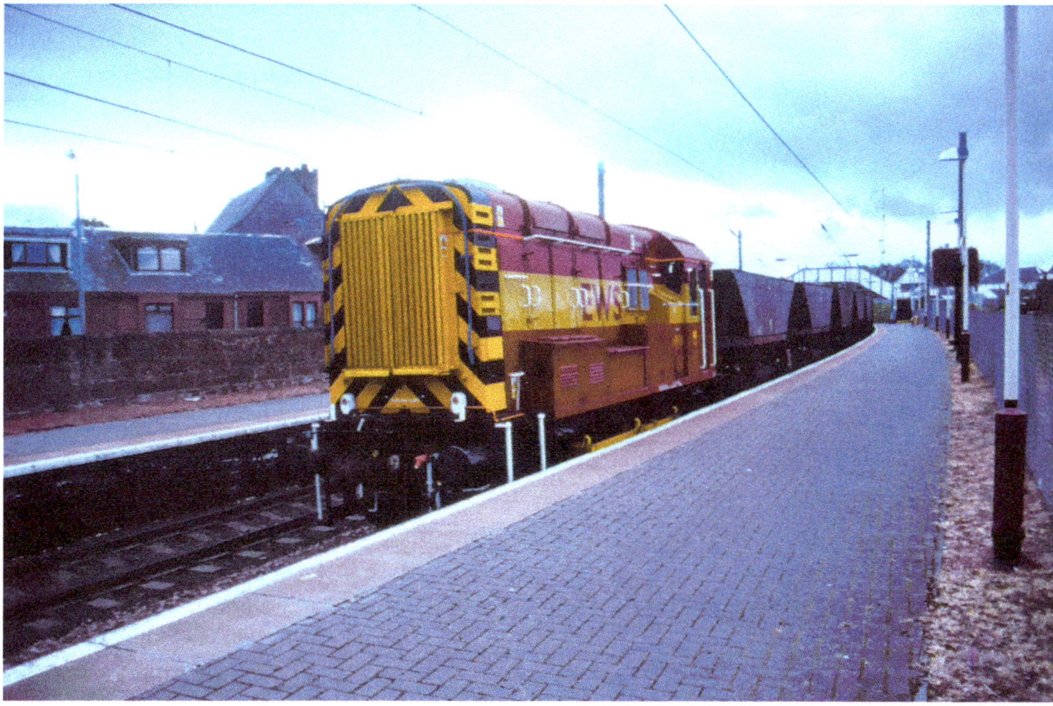

No. 08924 (ML) in EWS livery passes Newton-on-Ayr station with a transfer trip working from Ayr Depot to Falkland Yard. Taken June 2002.

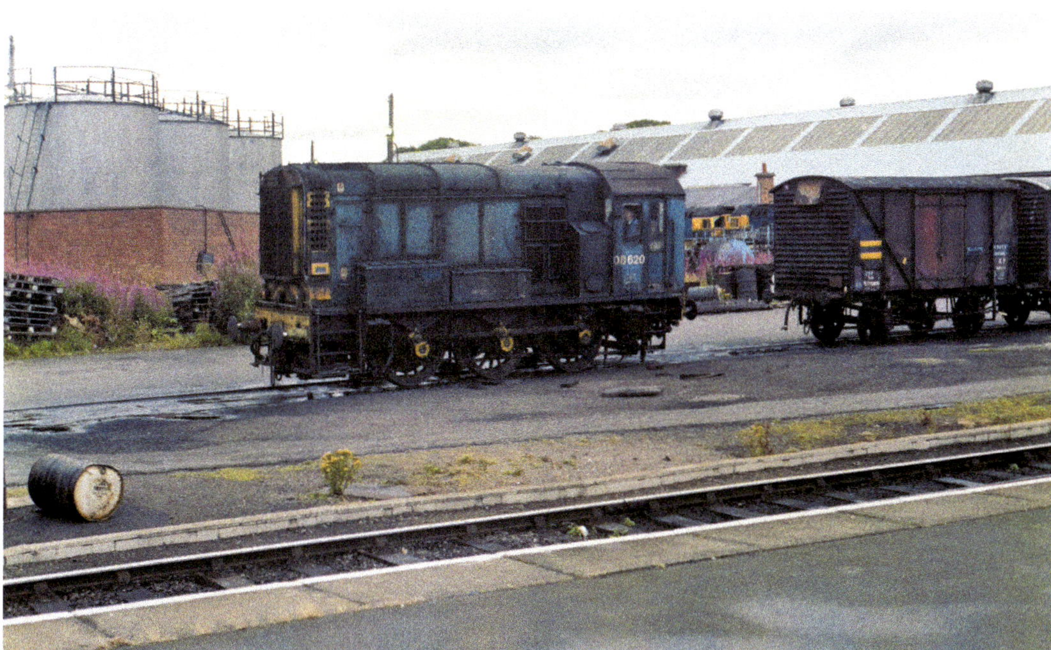

No. 08620 (IS) stabled at Inverness Depot. This loco was later converted to a Class 09 and renumbered 09205. Taken August 1981.

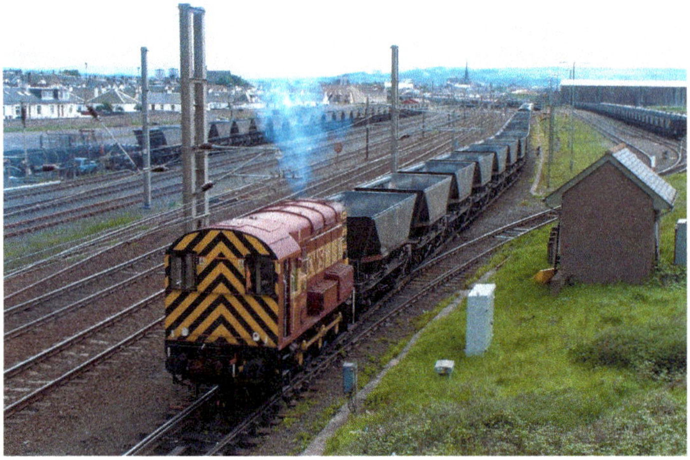

No. 08441 (ML) shunting empty MGR wagons at Falkland Yard. Taken August 2003.

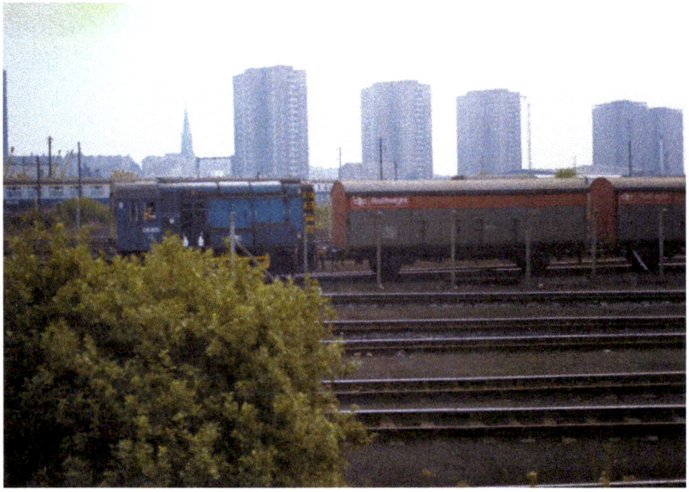

Left: No. 08851 (ED) shunting wagons in Springburn Works, Glasgow. Taken August 1981.

Below: No. 08730 (AY) heads light diesel at Hunterston Junction onto the Low Level branch to pick up coal wagons. Taken November 1980.

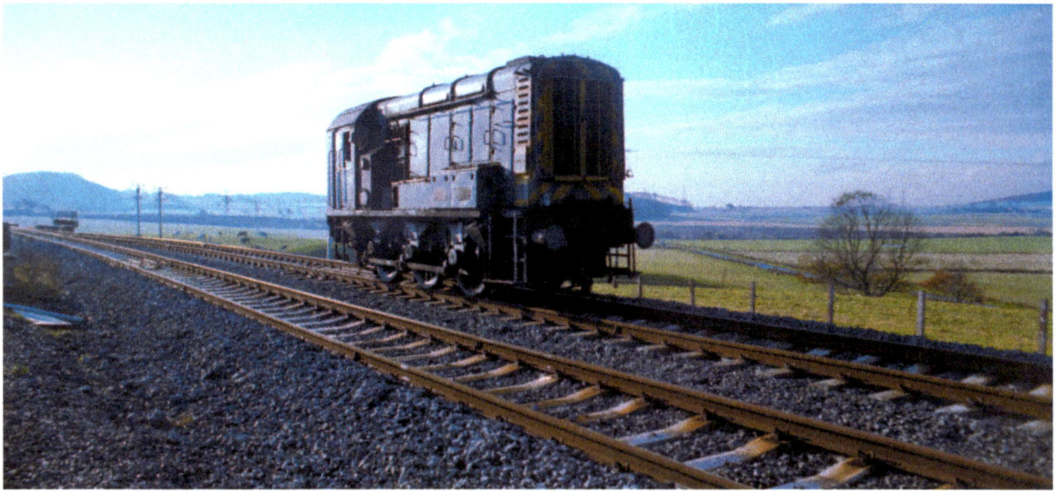

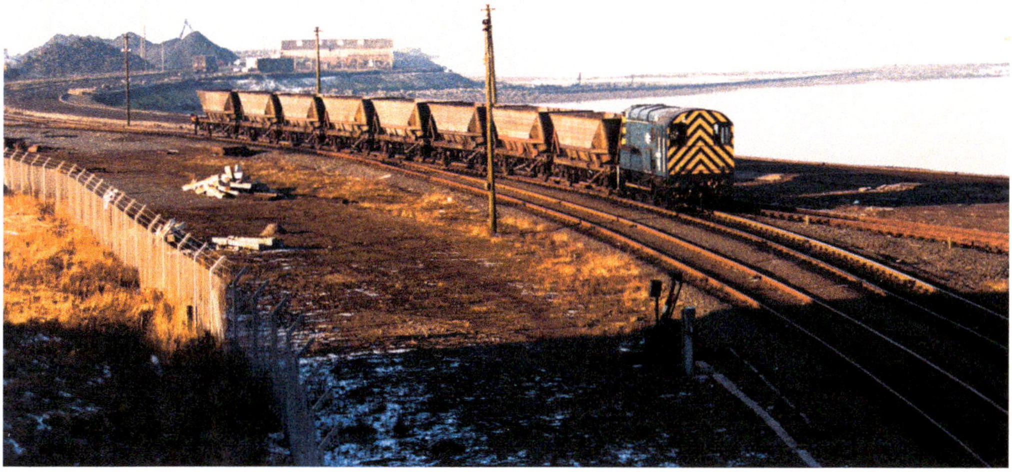

Above: No. 08675 (AY) at Ayr Harbour with empty MGR wagons returning from the disposal point to Falkland Yard. Taken February 1987.

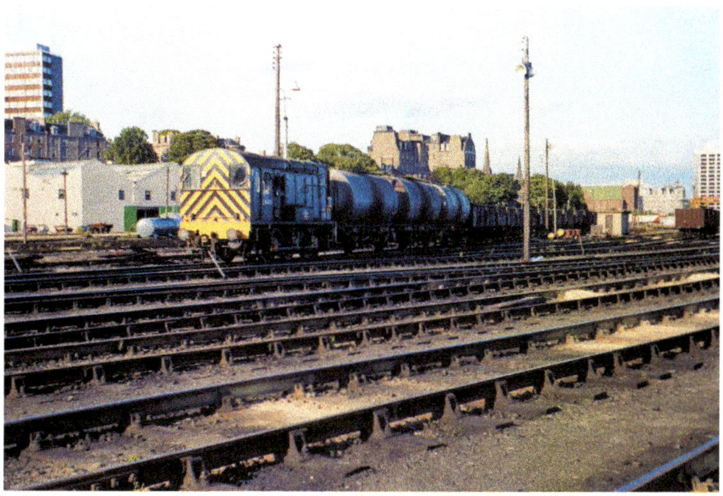

Right: No. 08424 (DE) with local trip working in Dundee Yard. Taken August 1981.

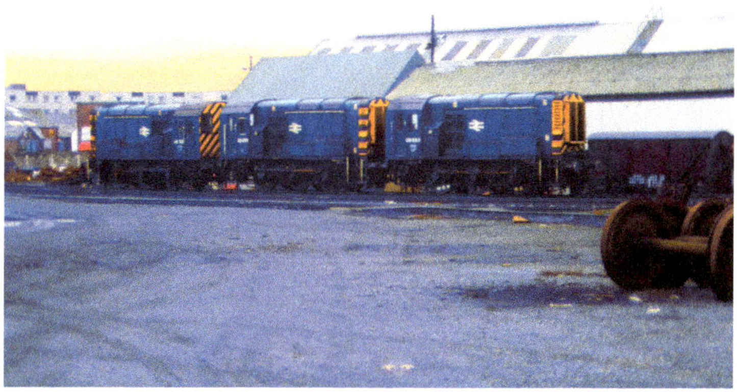

From left to right: Nos 08727, 08675 and 08693 (all Ayr) resting between shunting duties at Ayr Depot. Taken March 1988.

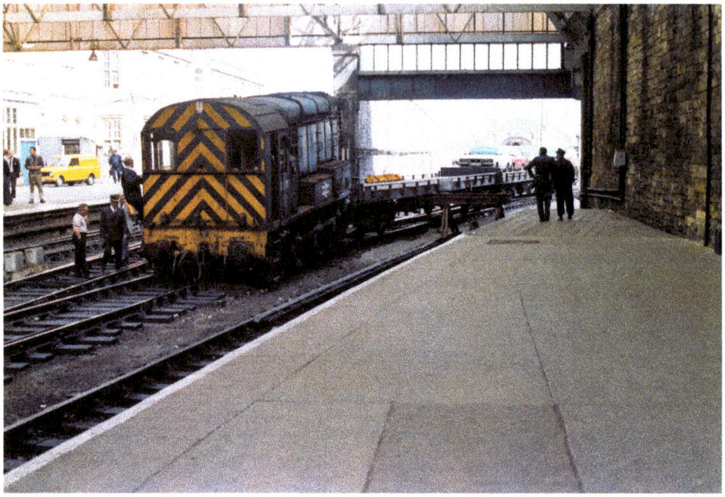

No. 08573 (DE) off the road. The loco had been derailed at a set of catch points whilst shunting Motorail wagons. Taken August 1981.

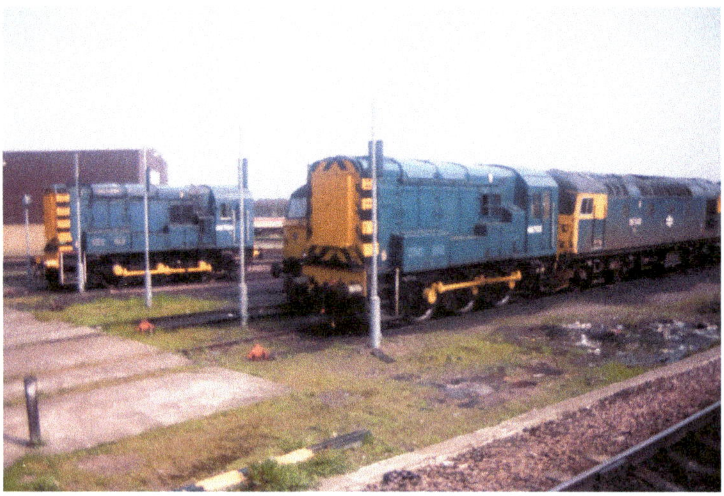

Nos 08764 and 08762 (both IS) stabled at Inverness Depot. Taken July 1984.

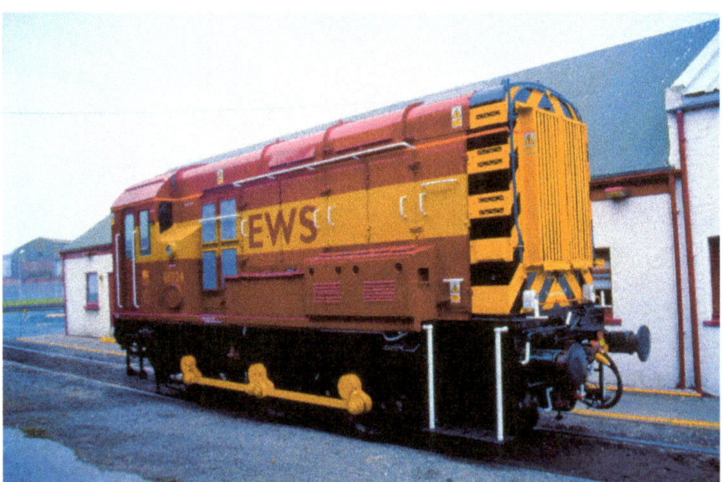

No. 08924 (ML) stabled at Ayr Depot. Taken May 2002.

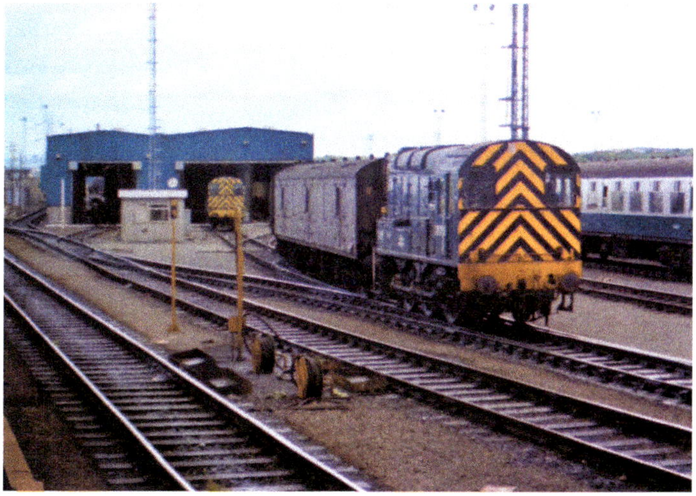

No. 08763 (HA) with a mixture of coaches and a newspaper van at Craigentinny Depot, Edinburgh. Taken August 1981.

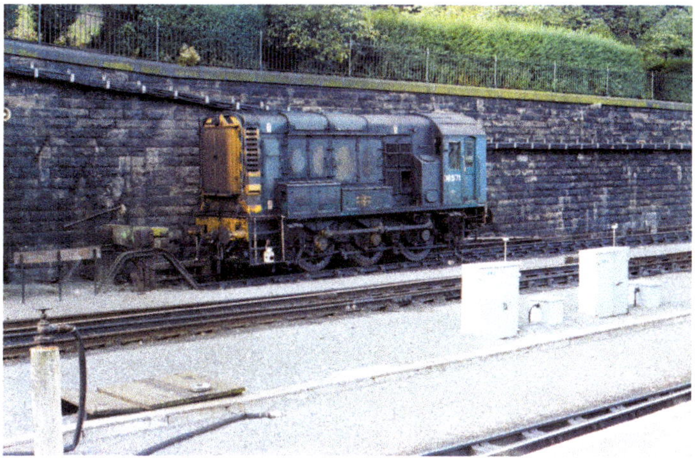

No. 08571 (HA) stabled between duties at Edinburgh Waverley. Taken August 1981.

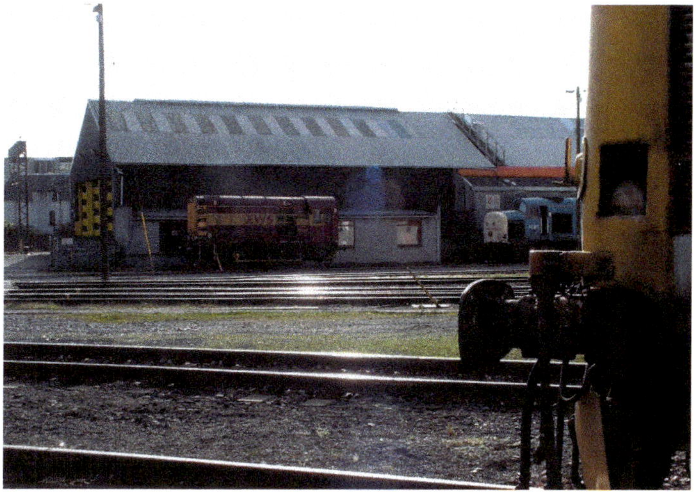

No. 08441 (ML) at Ayr Depot. In the background were withdrawn Class 37s that all made it into preservation and further use. Taken April 2002.

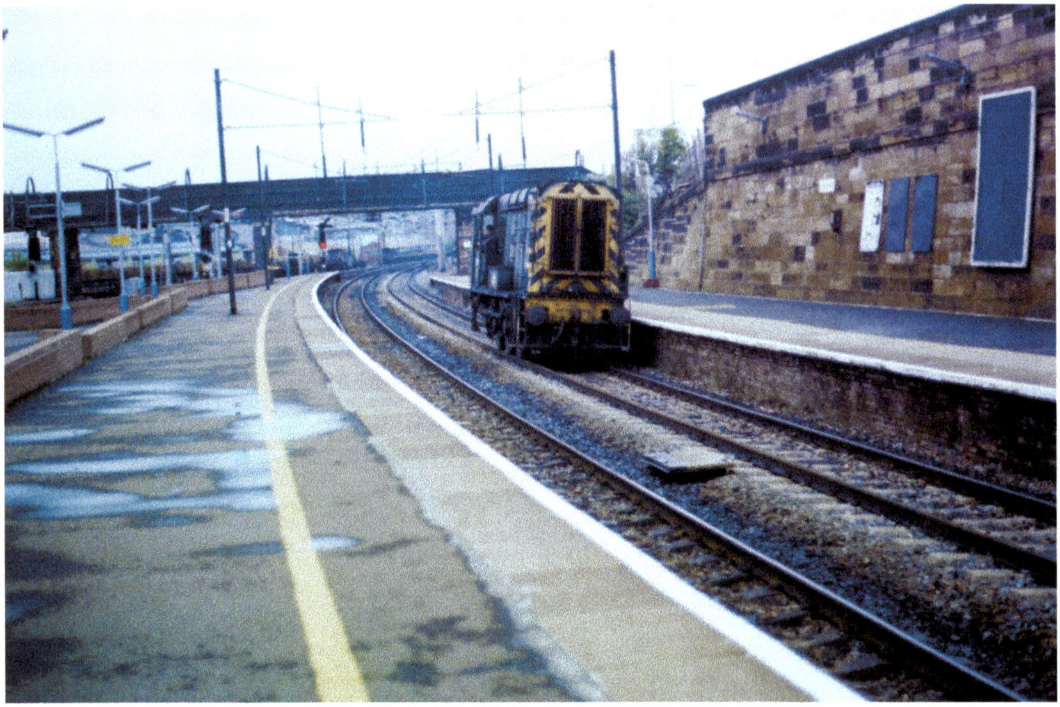

No. 08321 (ML) at Motherwell station heading from Motherwell Depot to the nearby Viewpark Sidings. Taken August 1981.

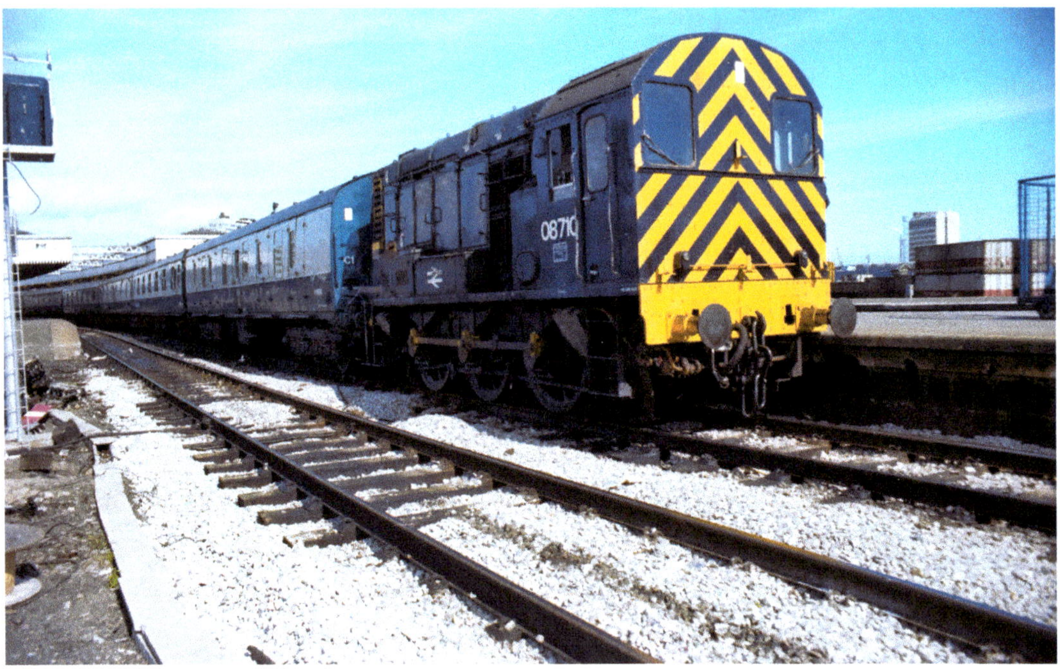

No. 08710 (AB) at Aberdeen ready to depart with ECS off the overnight Sleeper to Clayhills Depot for cleaning and maintenance. Taken August 1981.

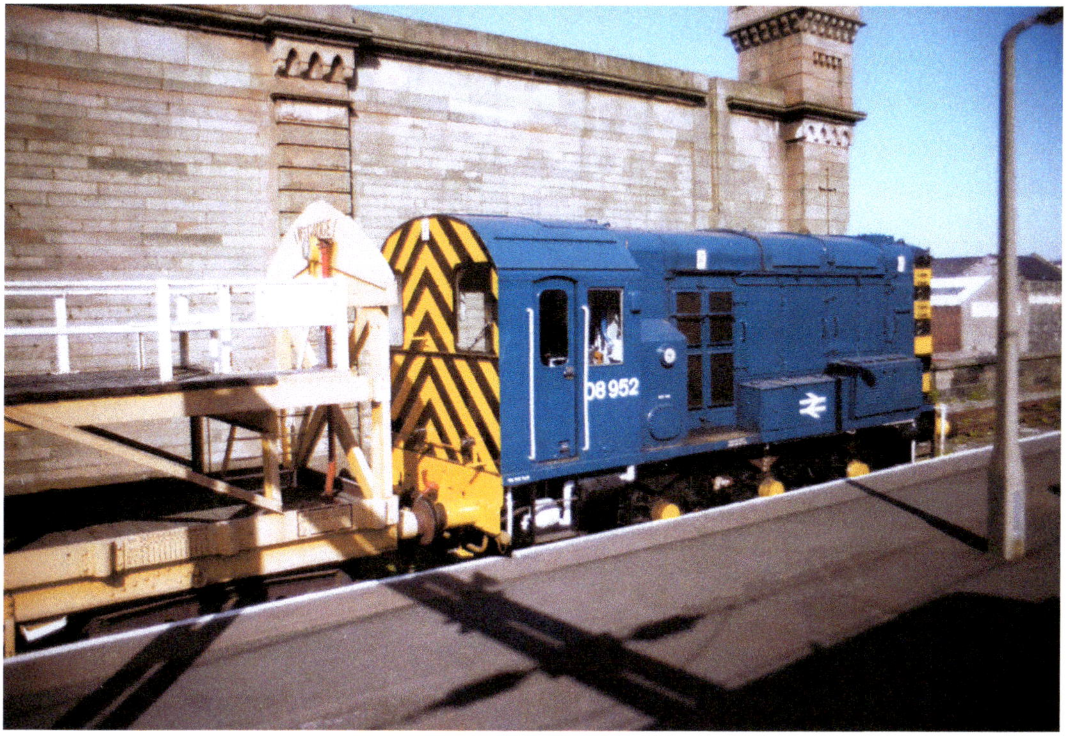

No. 08952 (ED) at Greenock Central with an engineers' wagon. Taken June 1984.

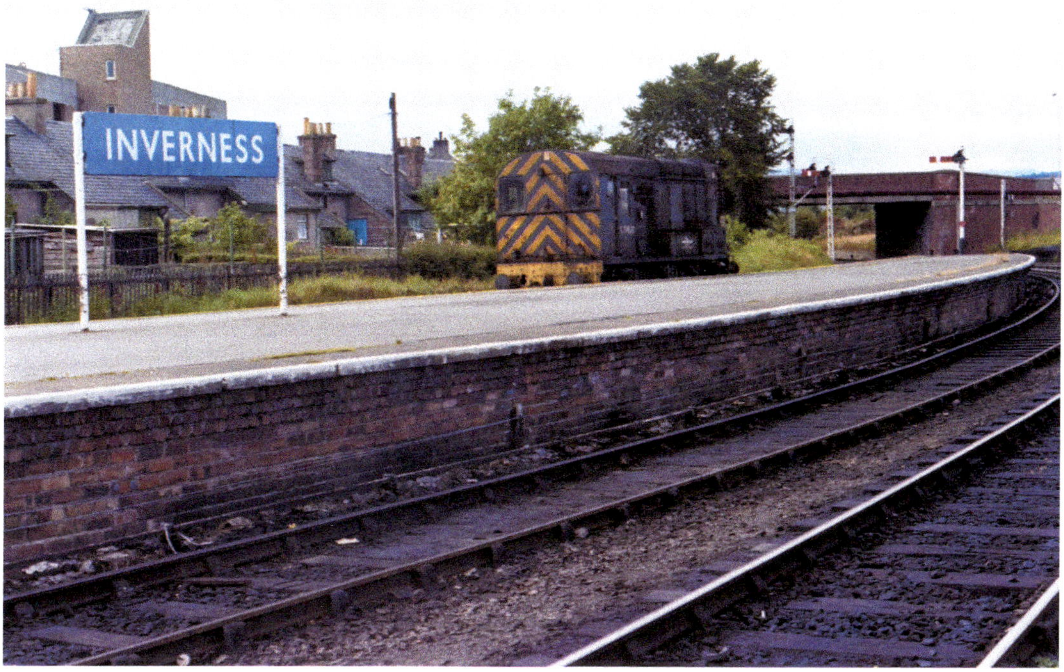

No. 08620 (IS) at Platform 1 of Inverness station. Taken August 1981.

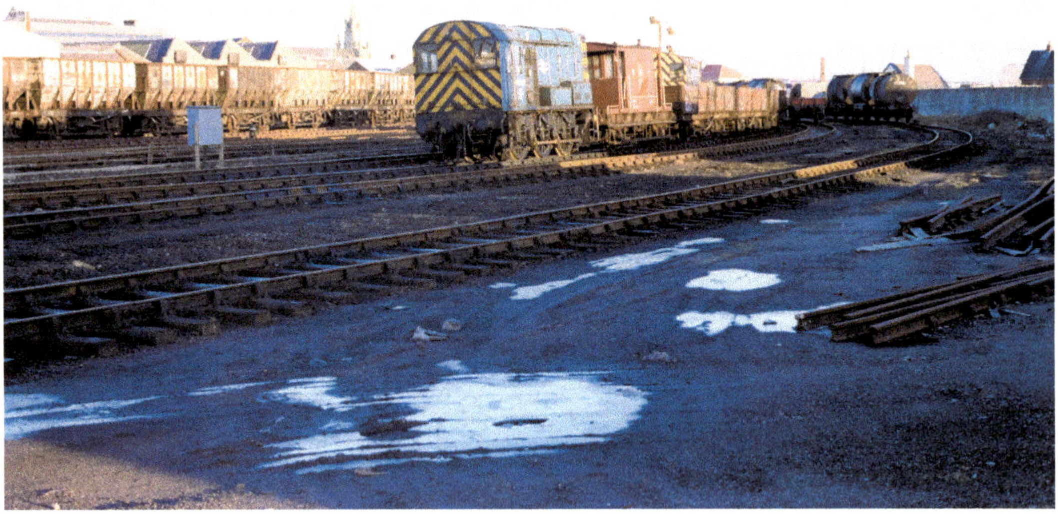

No. 08827 (AY) with wagons and tanks at Ardrossan Harbour. Taken December 1981.

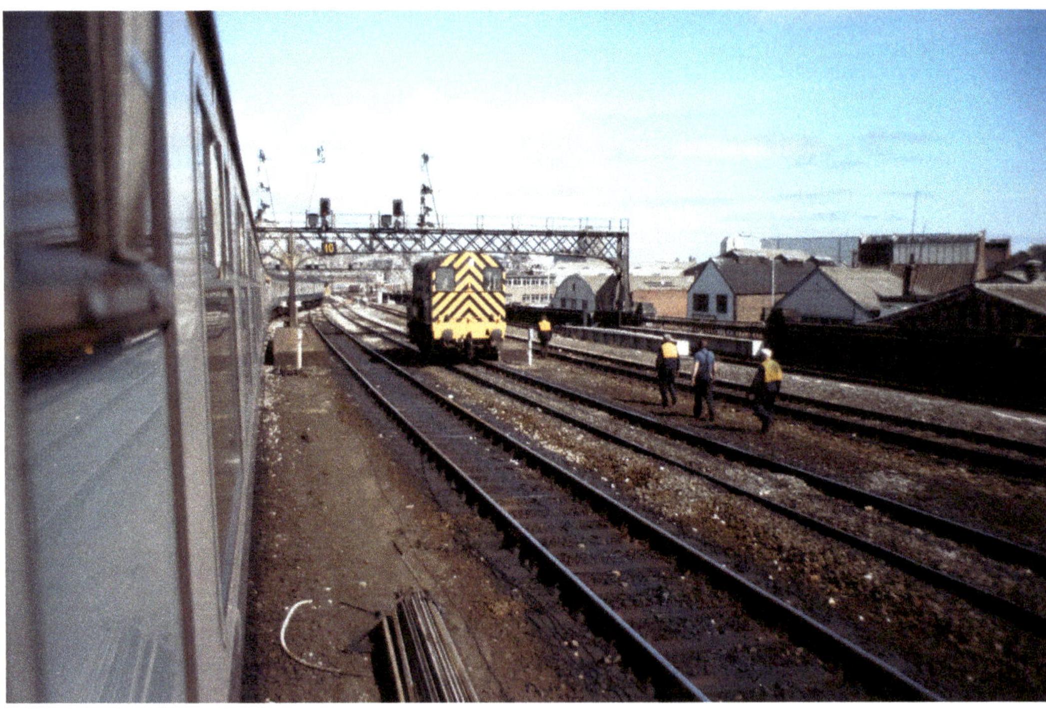

No. 08710 (AB) waits outside Aberdeen station for its next move with ECS. Taken August 1981.

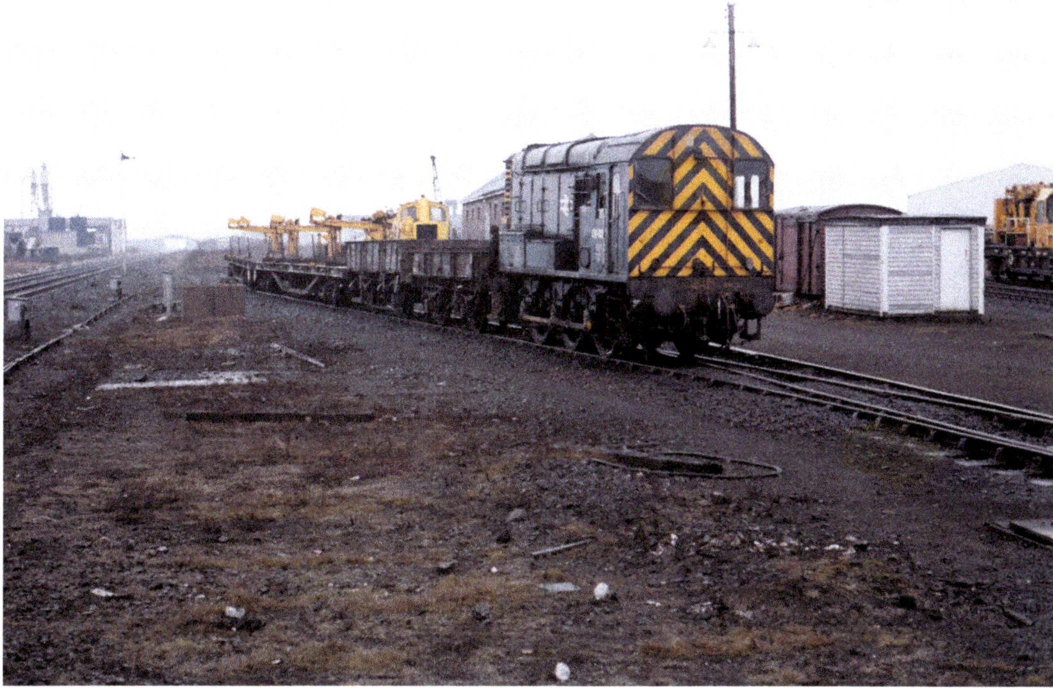

No. 8827 (AY) at Irvine West Yard shunting engineers' wagons. Taken December 1981.

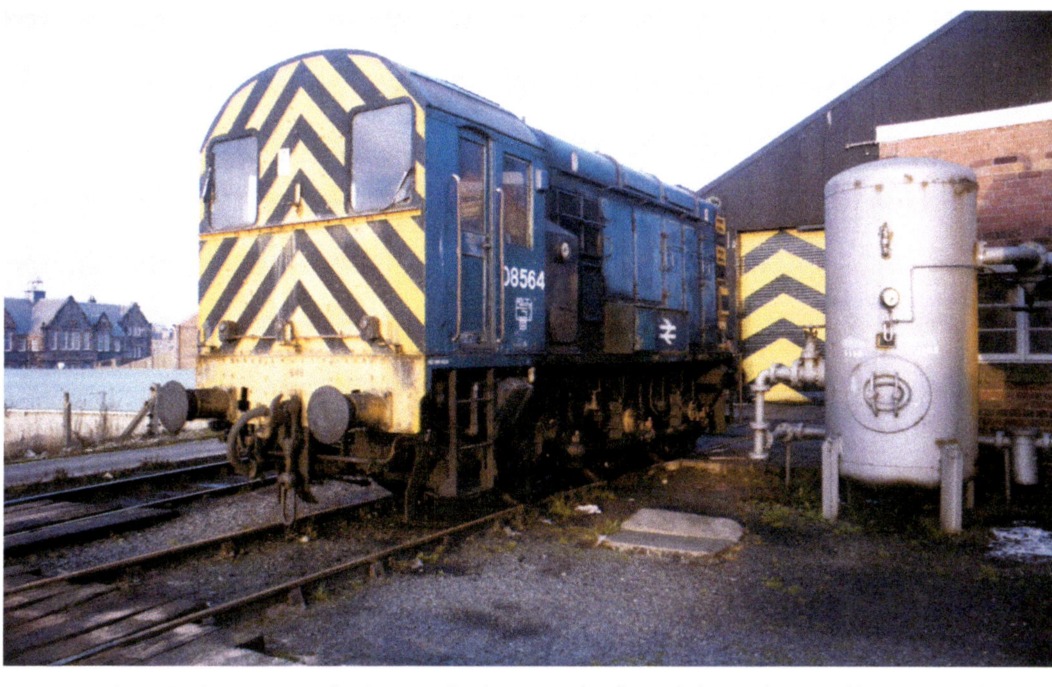

No. 08564 (HA) at Haymarket TMD. This loco was finally withdrawn from traffic in November 1985 and scrapped at J. McWilliams, Shettleston, by March 1986. Taken February 1985.

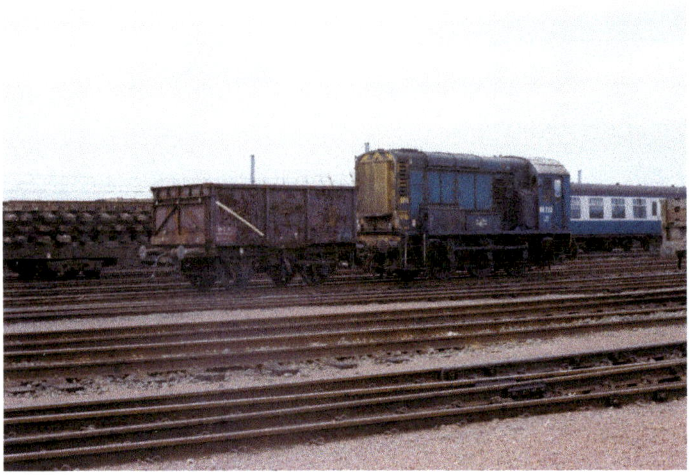

No. 08723 (ML) with a mineral wagon at Mossend Yard, near Bellshill. Taken May 1982.

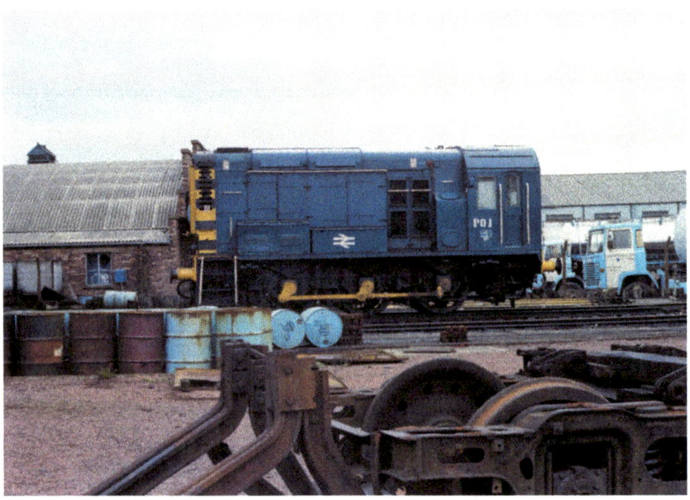

No. 08173 (ML), also numbered PO1, at Polmadie Depot. This was primarily the Polmadie pilot loco until withdrawn from traffic in January 1987 and finally scrapped by March 1987. Taken April 1982.

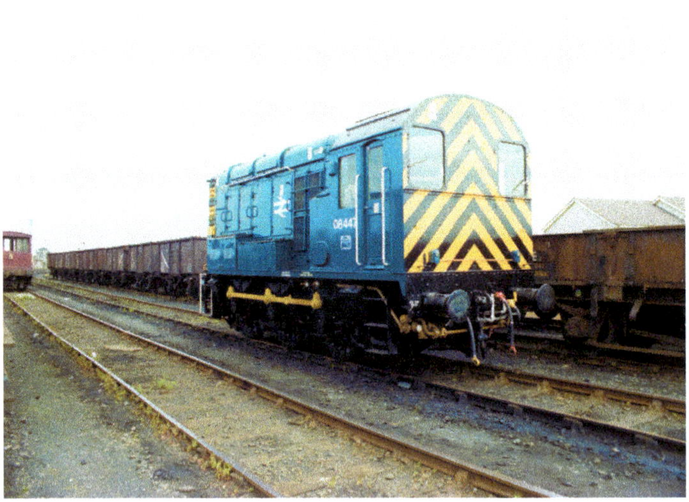

No. 08447 (AY) at Falkland Yard between shunting duties. This loco was withdrawn from traffic in November 1994 and later sold to Deanside Transit at Hillington in Glasgow in January 1995. Taken August 1983.

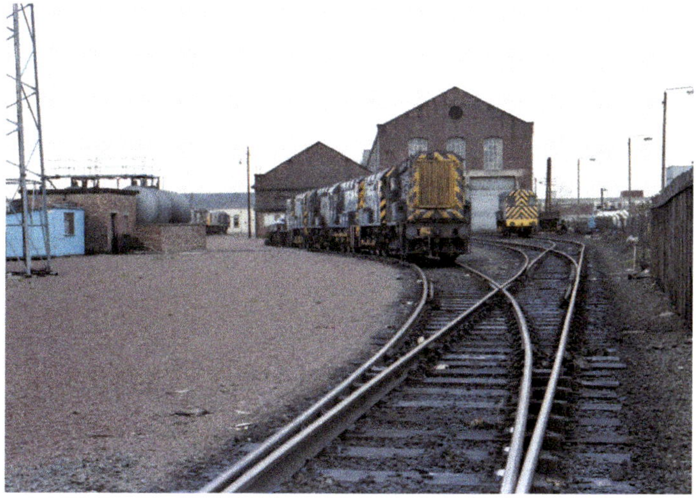

A collection of Class 08s at Polmadie Depot. From left to right: Nos 08727/736/621/859 (all ML) and 06008 (HQ), withdrawn from traffic, all stabled. Taken April 1982.

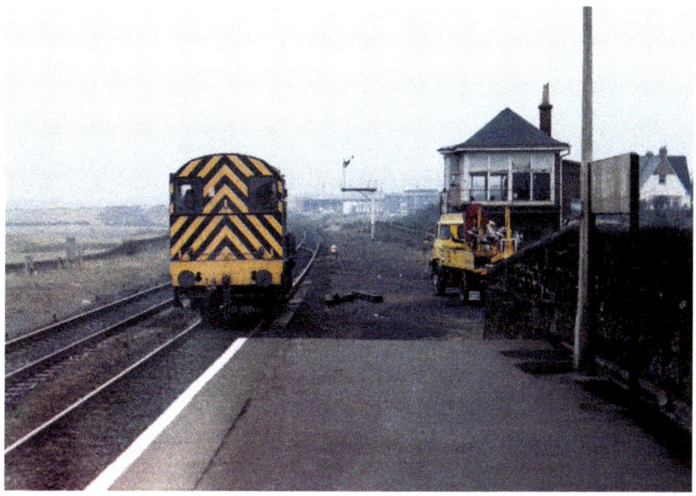

An unidentified 08 passes Prestwick signal box en route from Glengarnock steelworks to Ayr Depot for refuelling. Taken November 1980.

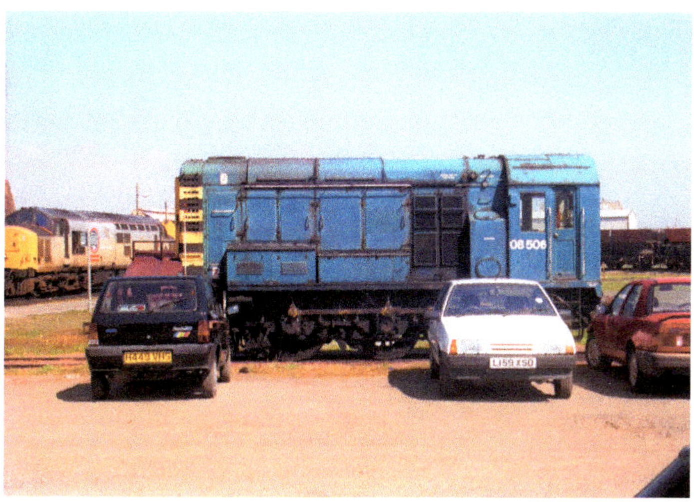

No. 8506 (AY) at Ayr Depot along with staff cars. All now history. Taken June 1997.

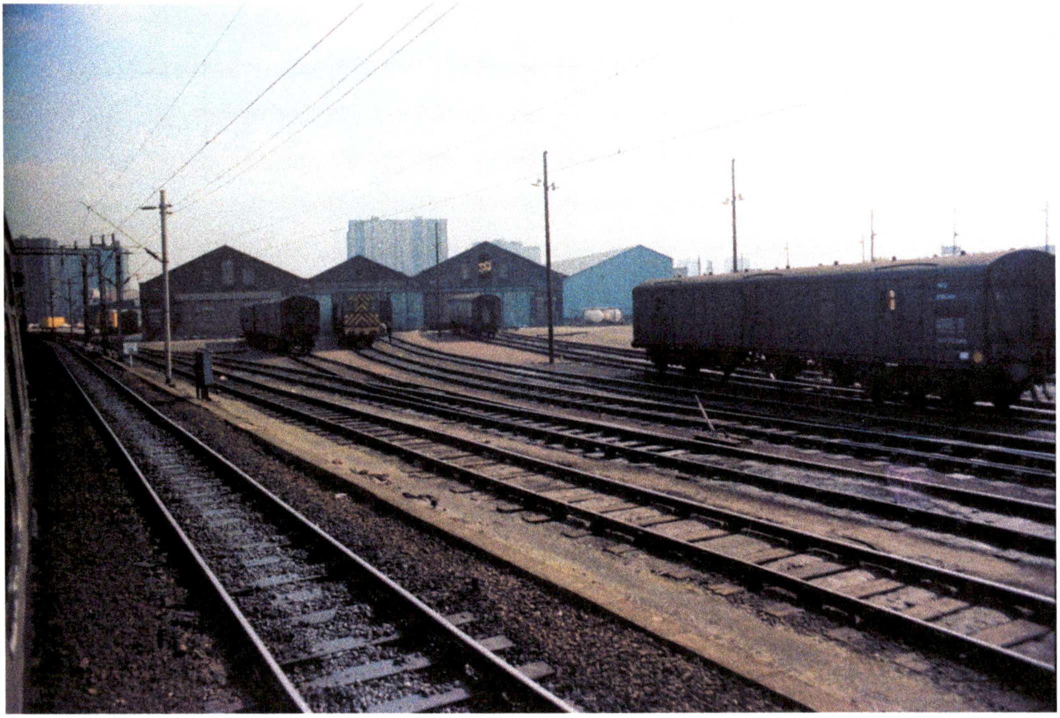

No. 08693 (ED) at Glasgow's Salkeld Street Parcels Depot. This yard closed in 1991. Taken August 1981.

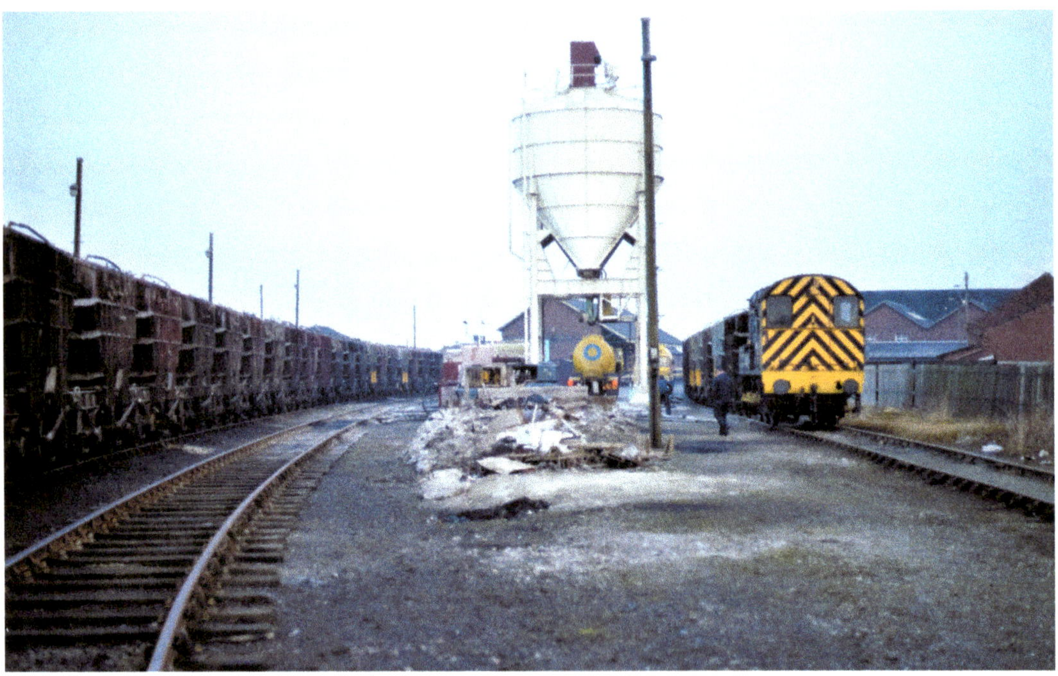

No. 08345 (AY) at Irvine East Yard with Blue Circle cement wagons. Taken March 1982.

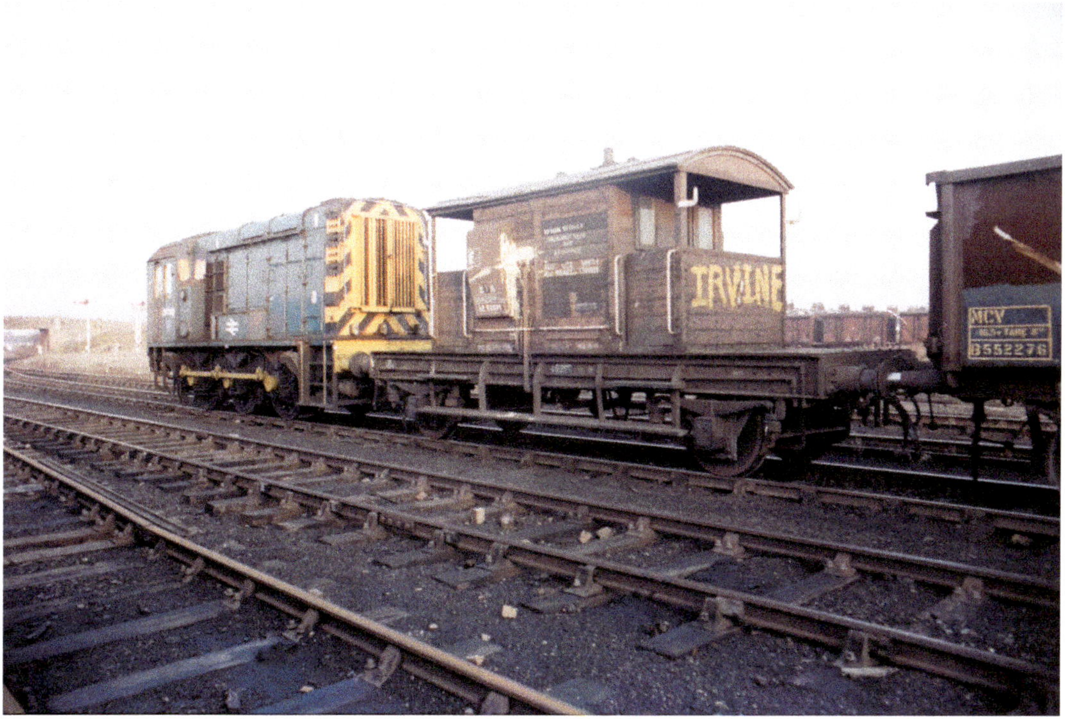

No. 08449 (AY) with a brake van and mineral wagon at Falkland Yard. Taken December 1982.

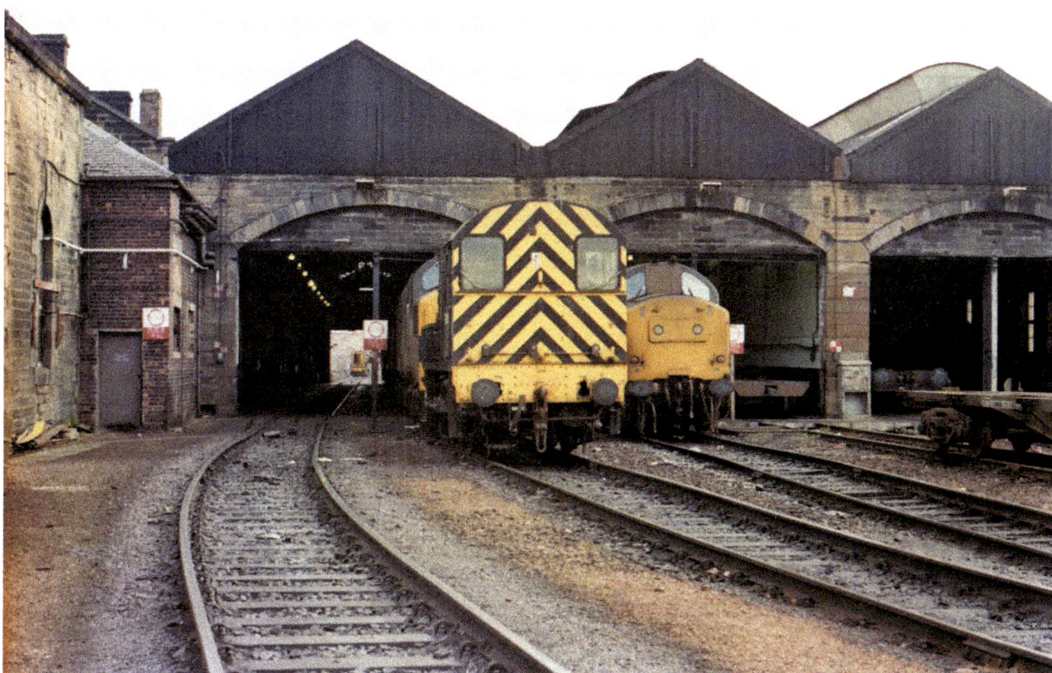

An unidentified 08 in company with two Class 37s at Motherwell Depot. Taken April 1982.

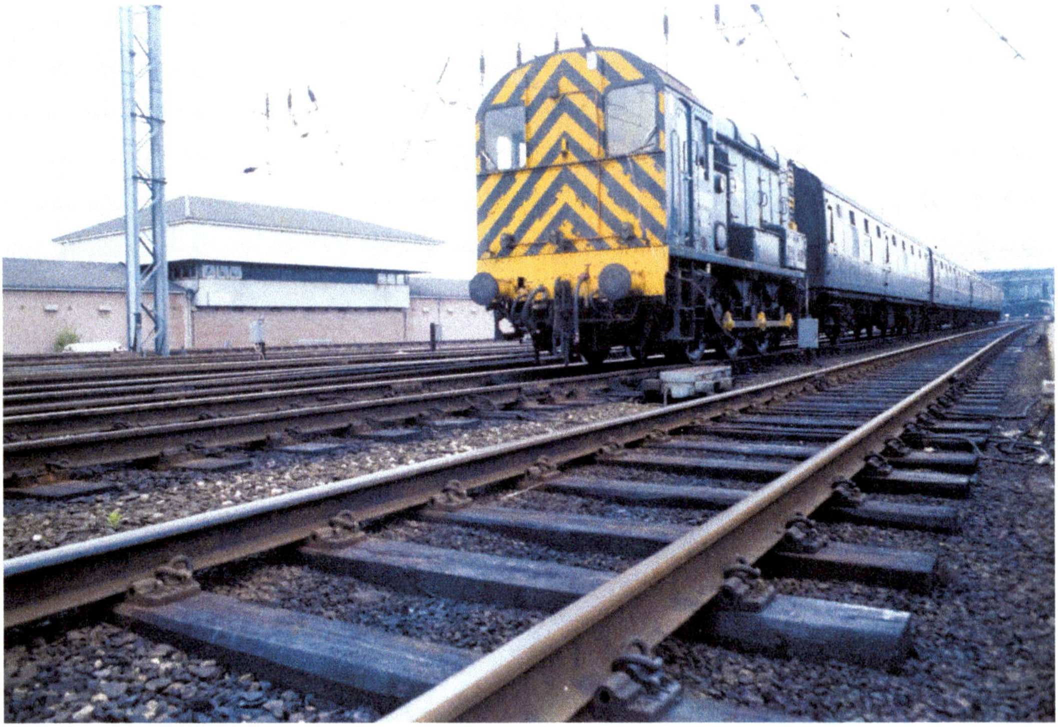

No. 08910 (KD) at the south end of Carlisle station shunting passenger ECS for their next working. Taken April 1983.

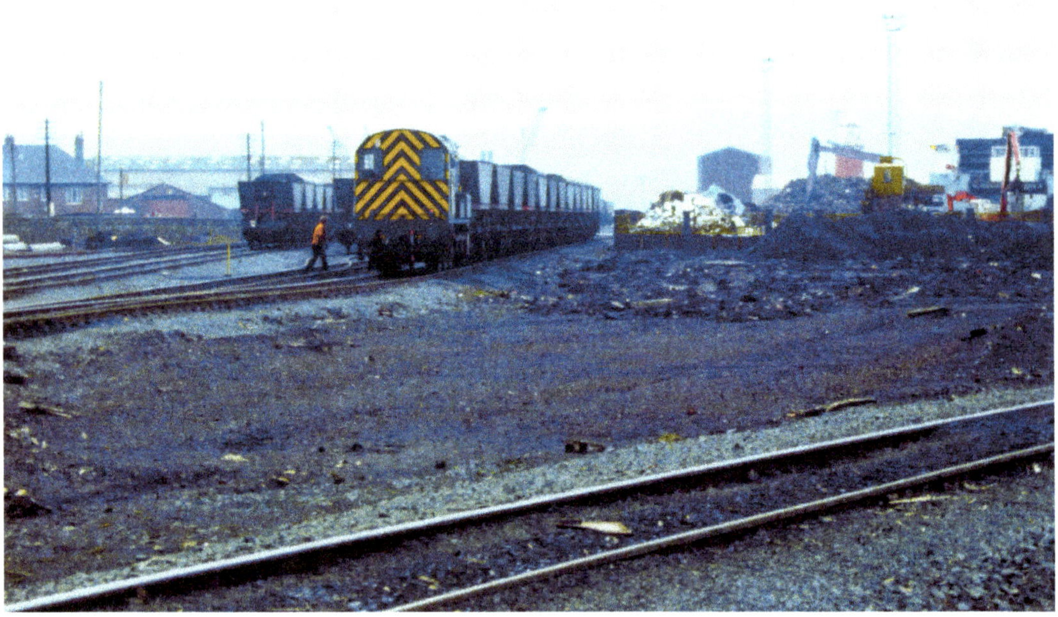

No. 08675 (AY) with MGR service at the coal disposal area at Ayr Harbour. Taken April 1987.

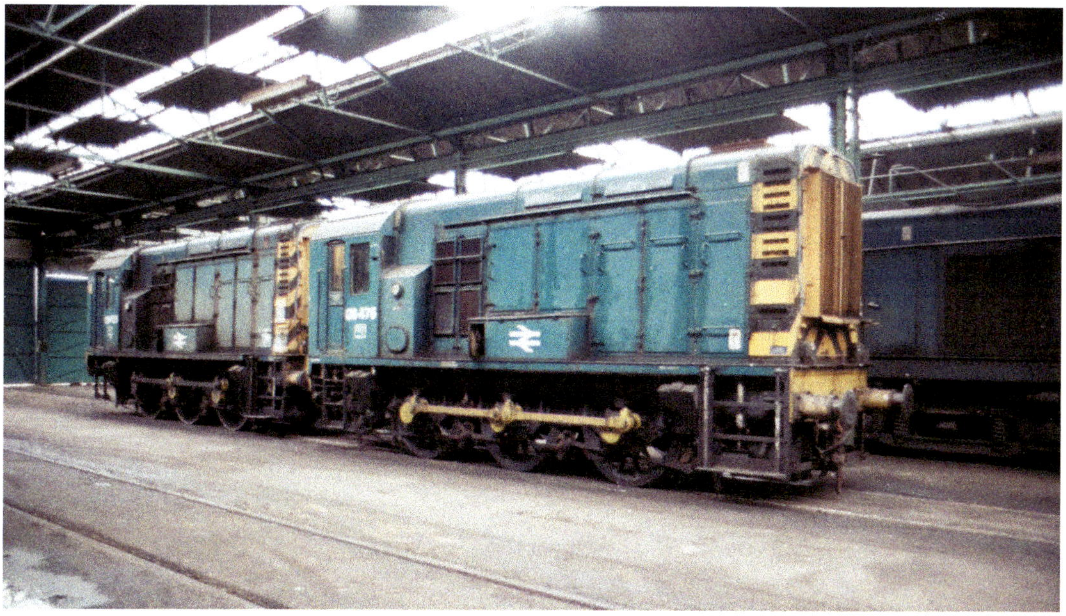

Nos 08425 and 08476 (both HQ) stabled inside Dunfermline Townhill Depot. No. 08425 was taken to Swindon Works and was broken up by August 1986. No. 08476 fared better. It was also moved to Swindon Works, but was bought by the Swanage Railway Preservation Society and was moved there in March 1986. It was also repainted into its original number – D3591. Taken April 1983.

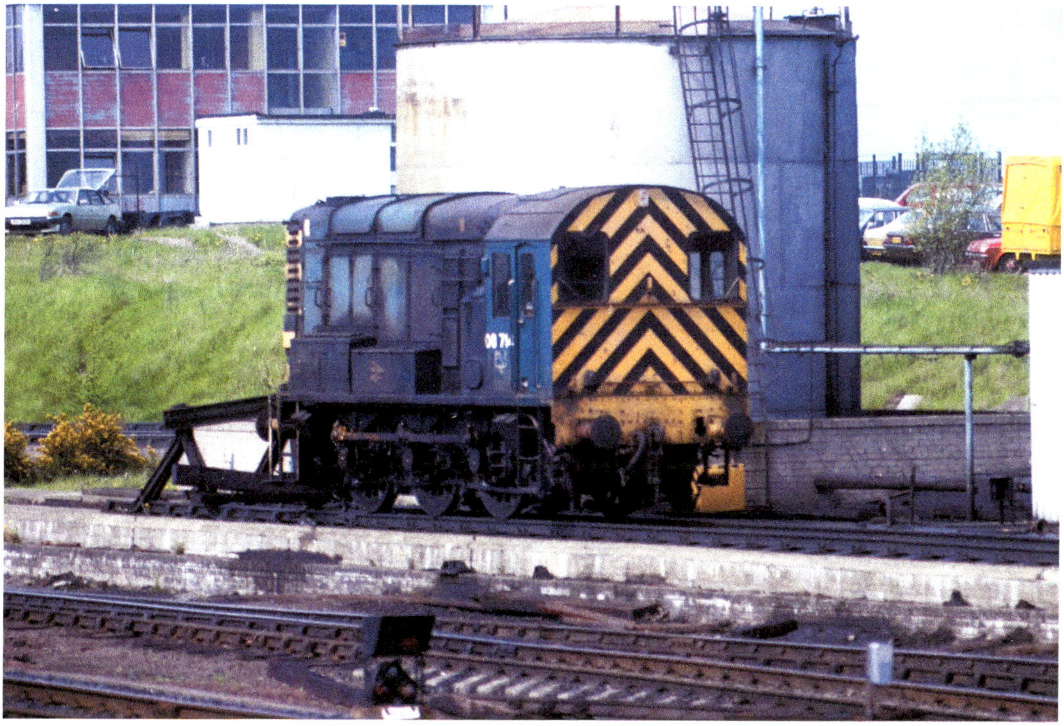

No. 08714 (HA) at Millerhill Yard in-between shunting duties. Taken April 1987.

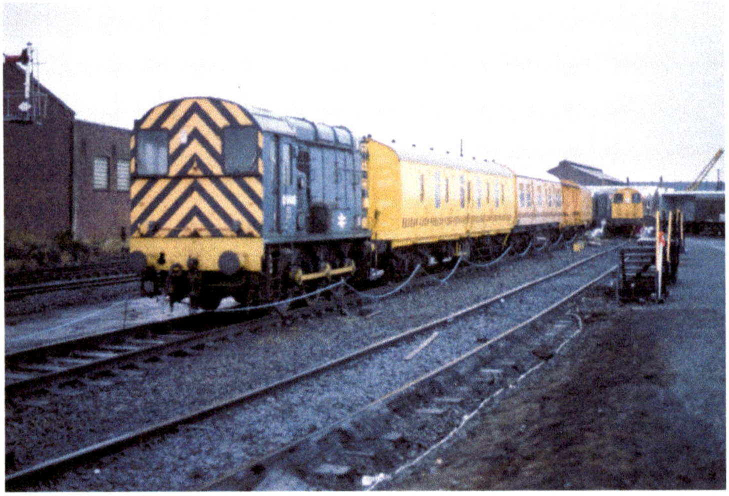

No. 08446 (AY) at Ayr Depot with engineers' breakdown coaches. Taken October 1983.

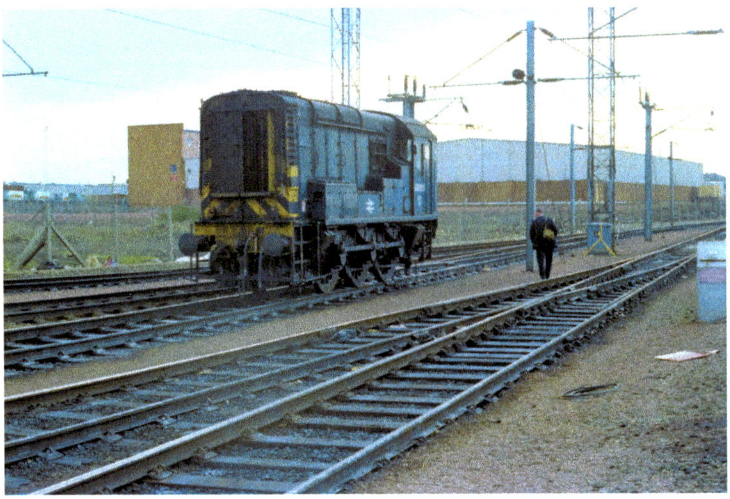

No. 08853 (ED) with driver in-between shunting duties at Polmadie Depot, Glasgow. Taken April 1982.

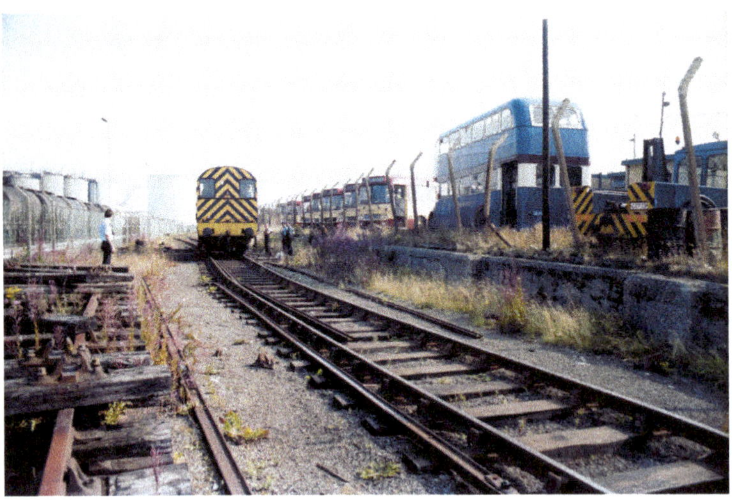

No. 08446 (AY) on the connecting line between the Shell Sidings and Ardrossan Harbour Exchange Sidings. Taken August 1983.

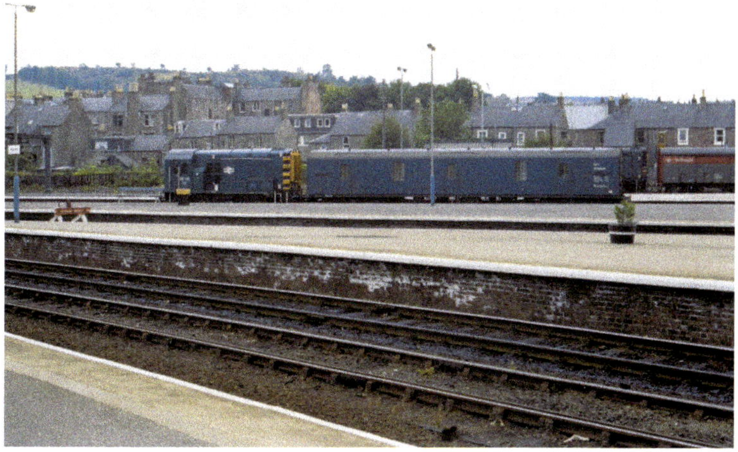

No. 08712 (TJ) shunting a General Utility Vehicle (GUV) coach at Perth station. Taken August 1984.

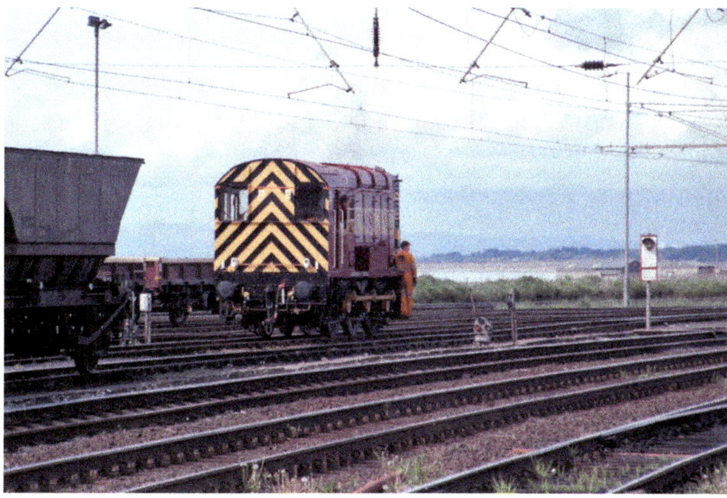

No. 08441 (ML) with a shunter on the side ladder at Falkland Yard. Taken May 2002.

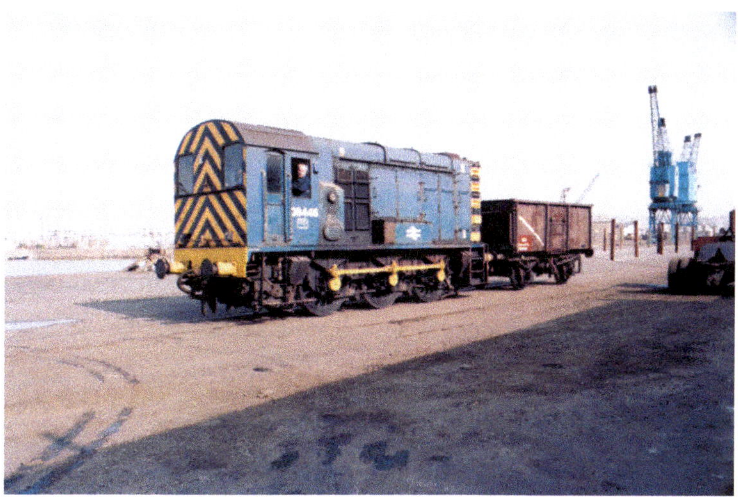

No. 08446 (AY) with a mineral wagon at Ardrossan Harbour. Taken August 1983.

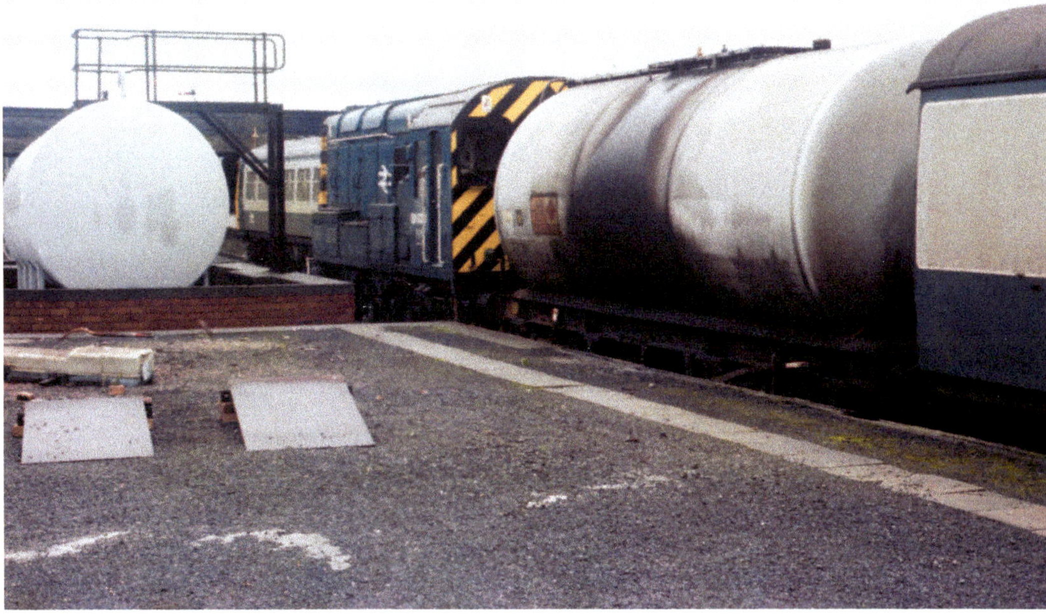

No. 08620 (ED) with a fuel tank wagon and coach stabled in the sidings at the north end of Stirling station. Taken October 1983.

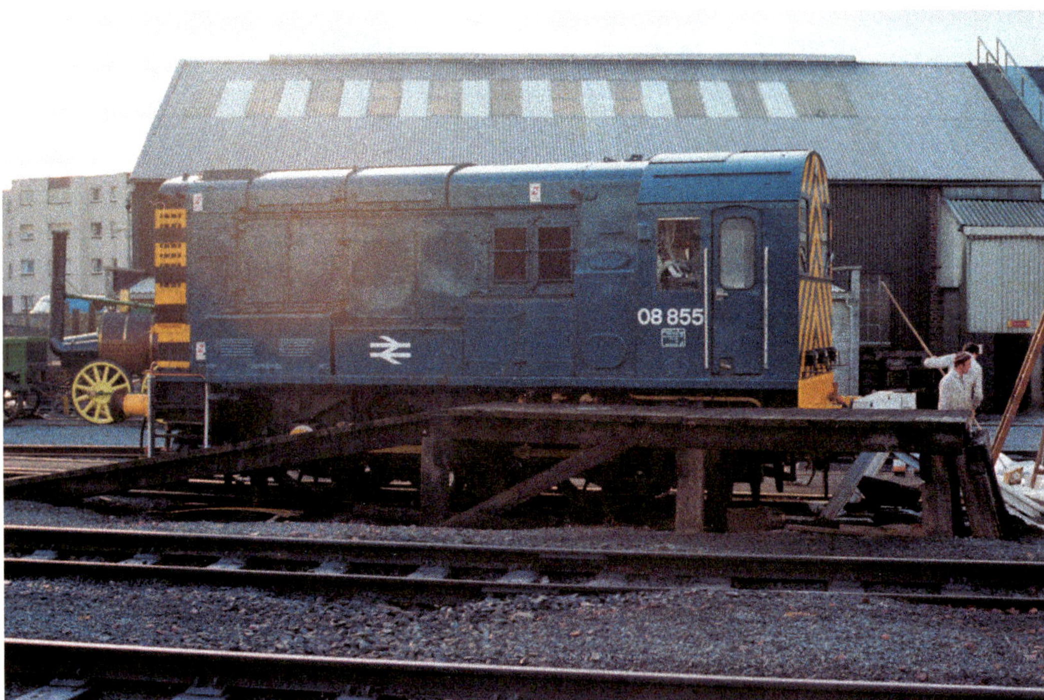

No. 08855 (AB) at Ayr Depot. The loco was on display for a BR open day. Taken October 1983.

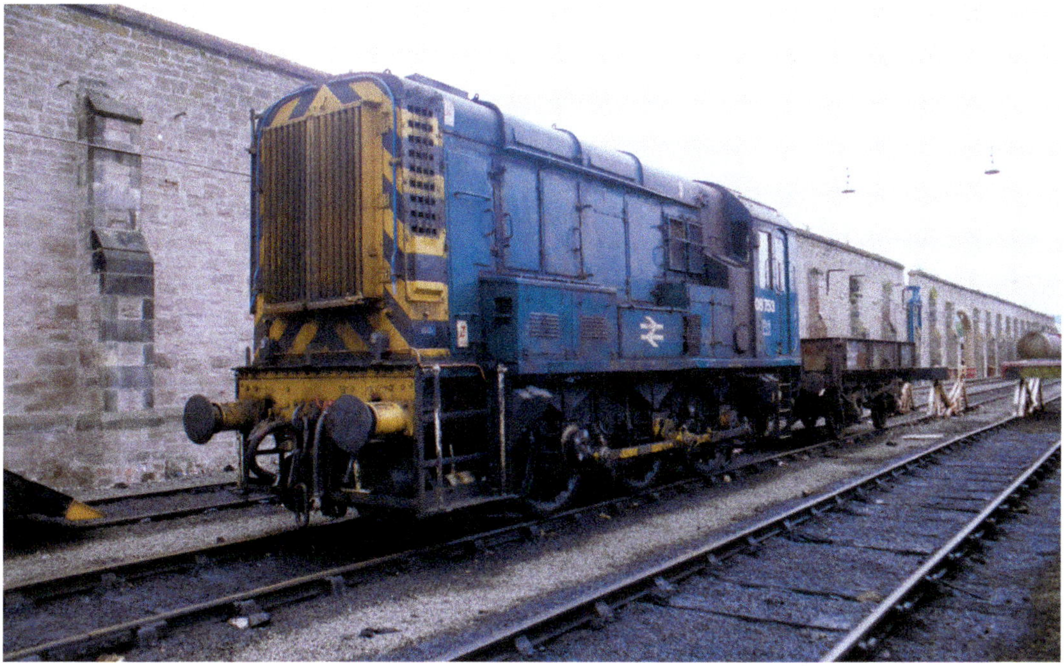

No. 08753 (DE) at Perth Sidings with an engineers' wagon. Taken August 1984.

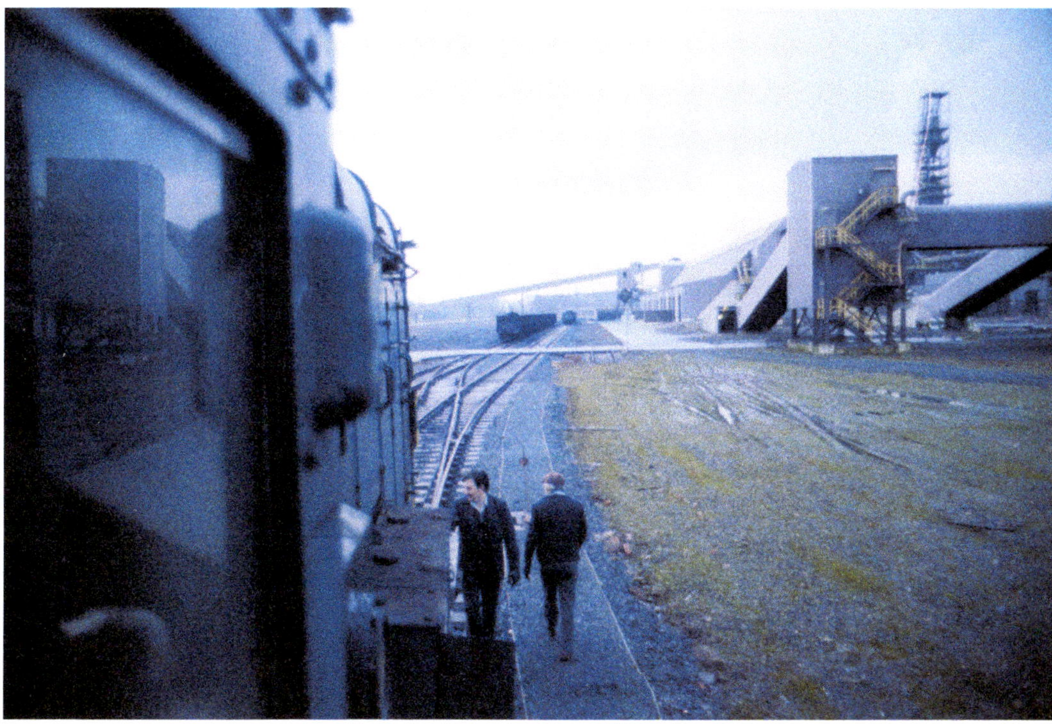

No. 08449 (AY) at Hunterston Low Level. Shunter and driver having discussed their next move. Taken December 1983.

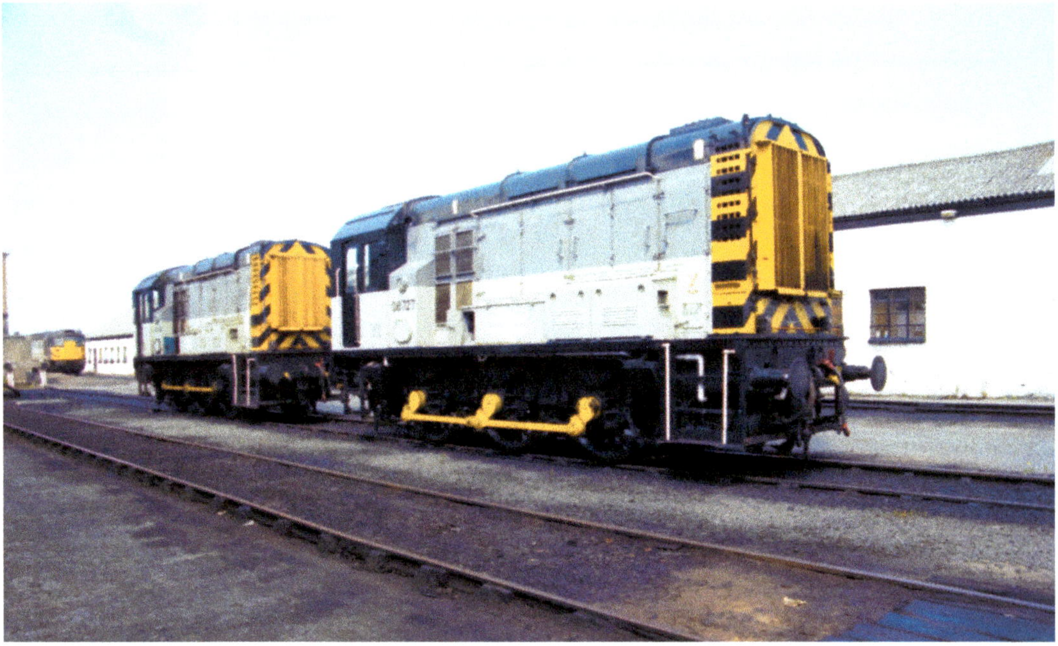

Nos 08727 and 08586 (both AY) in coal sector livery at Ayr Depot. No. 08727 was withdrawn in March 1993 and sold for scrap to MC Metal Processing at Springburn Works. It was cut up by September of the same year. No. 08586 was withdrawn from traffic in December 1995 and cut up on site at Ayr Depot before being finally dismantled by May 2000. Taken June 1990.

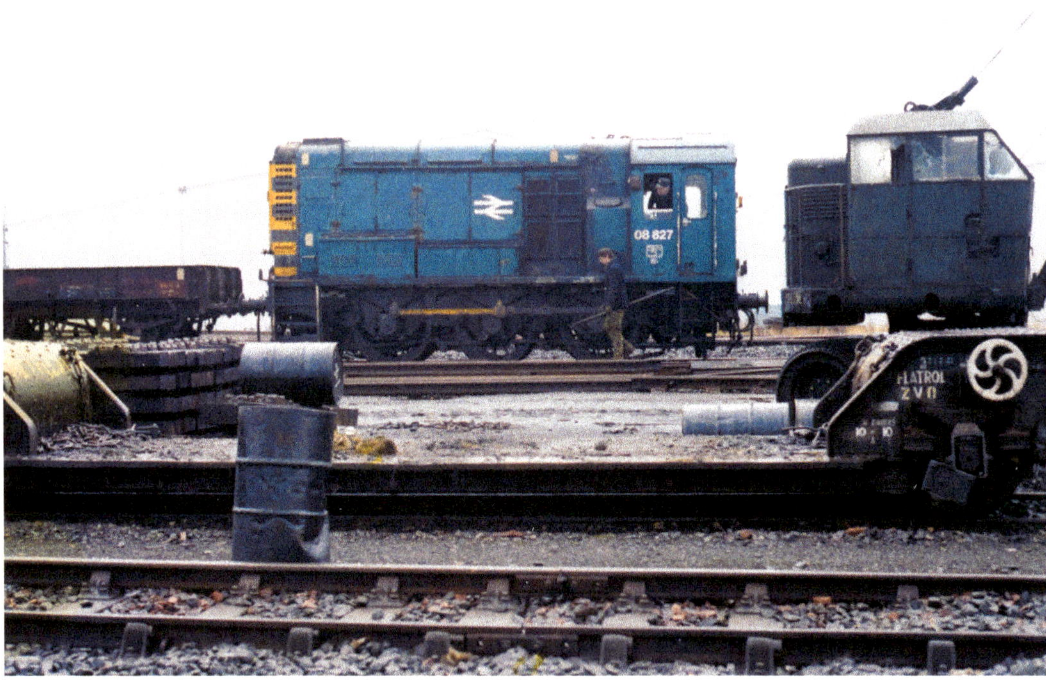

No. 08827 (AY) at Irvine West Yard shunting engineers' wagons. Taken December 1981.

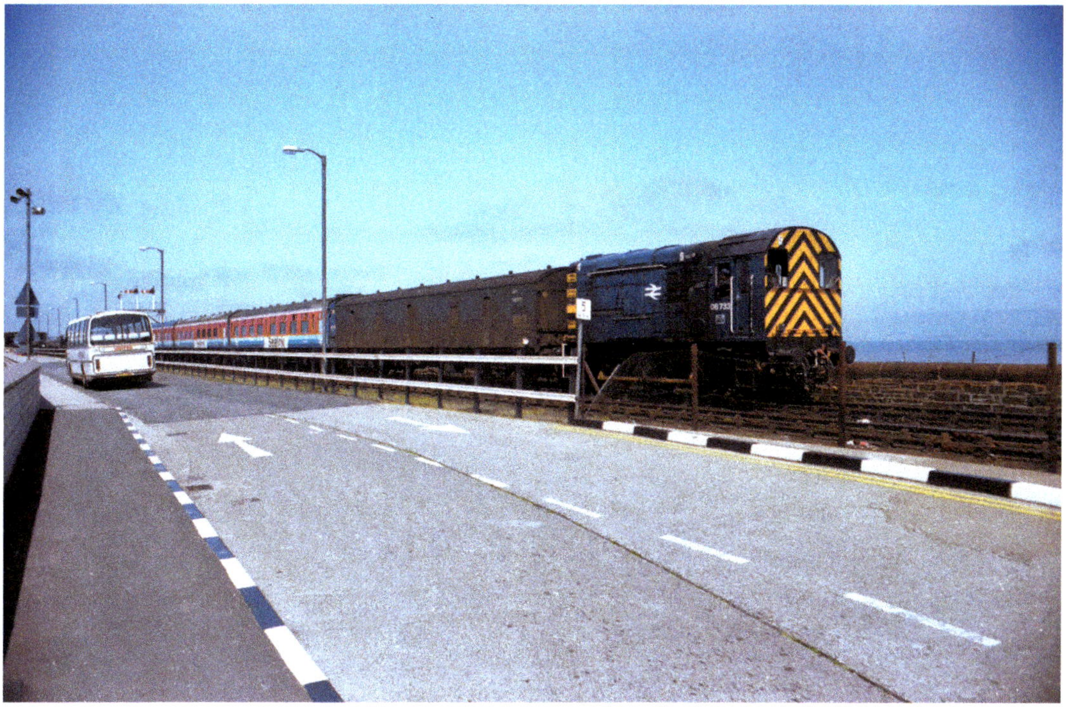

No. 08733 (AY) at Stranrear Harbour with Mk 1 Sealink stock and a GUV parcels van being shunted into the station. Taken May 1984.

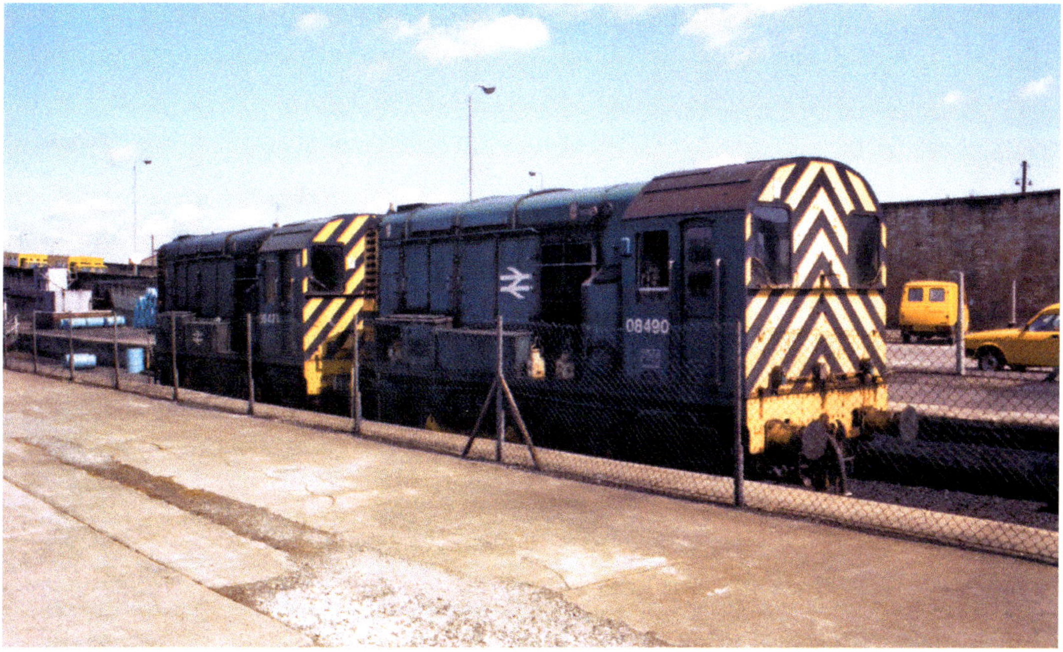

Nos 08424 and 08490 (both TJ) stabled between duties at the north end of Perth station. Taken August 1984.

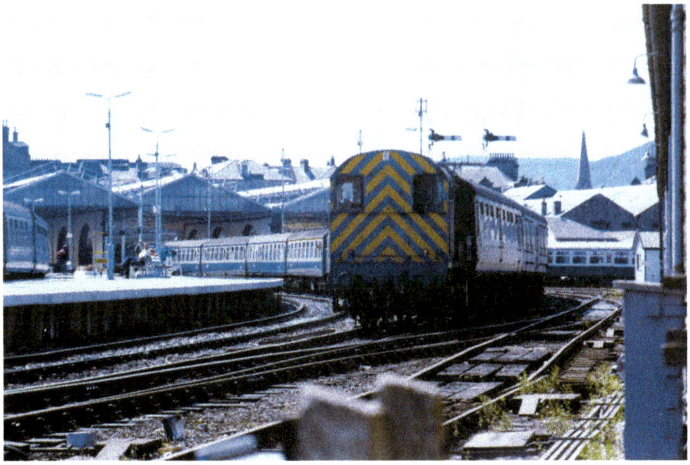

No. 08754 (IS) at Inverness station attaching to passenger stock for cleaning in the nearby depot road. Taken July 1987.

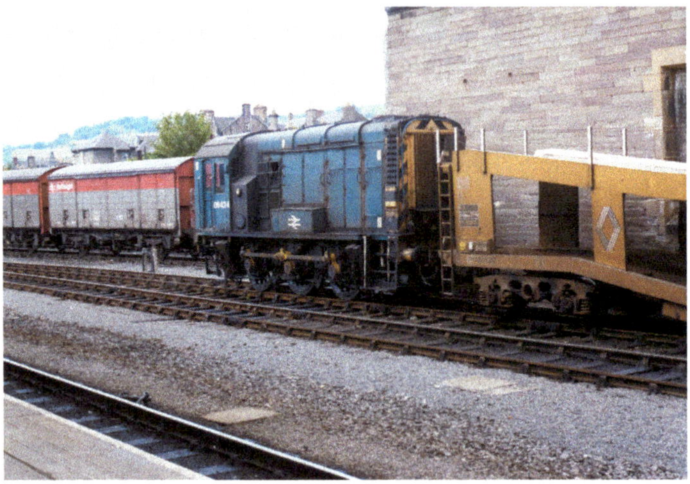

No. 08424 (TJ) at Perth with a Cartic wagon. These wagons were used to move cars and vans around the country. Taken August 1984.

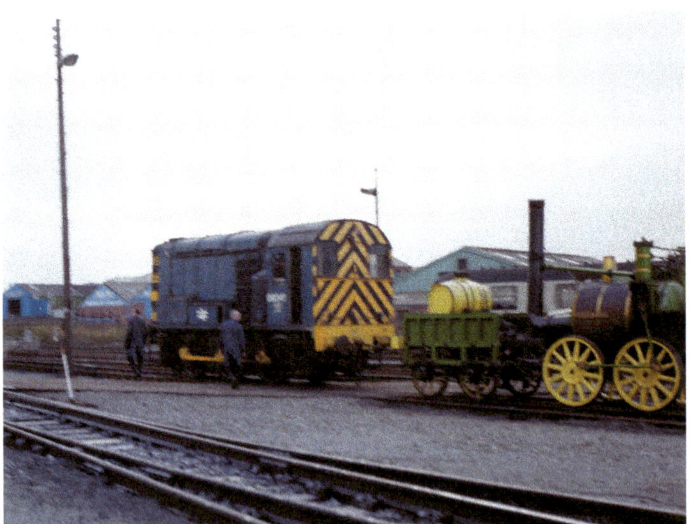

No. 08345 (AY) with steam replica loco *Sans Pareil* at Ayr Depot. Both locos took part in the BR open day. Taken October 1983.

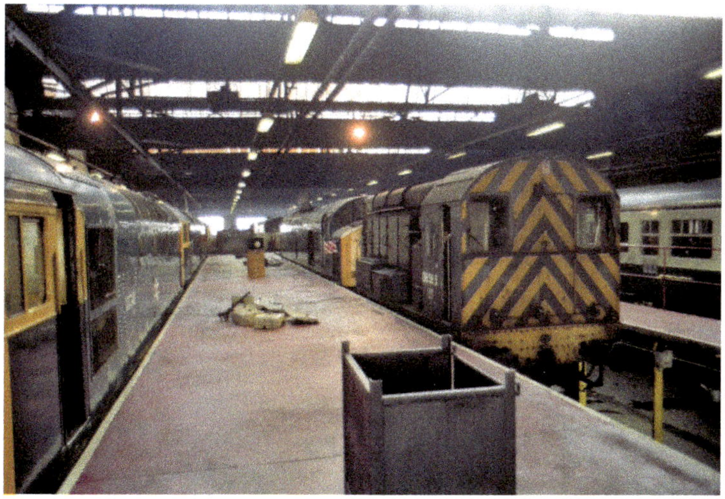

No. 08565 (HA) inside Haymarket TMD, with a Class 27, 40 and 101 power car for company. Taken August 1981.

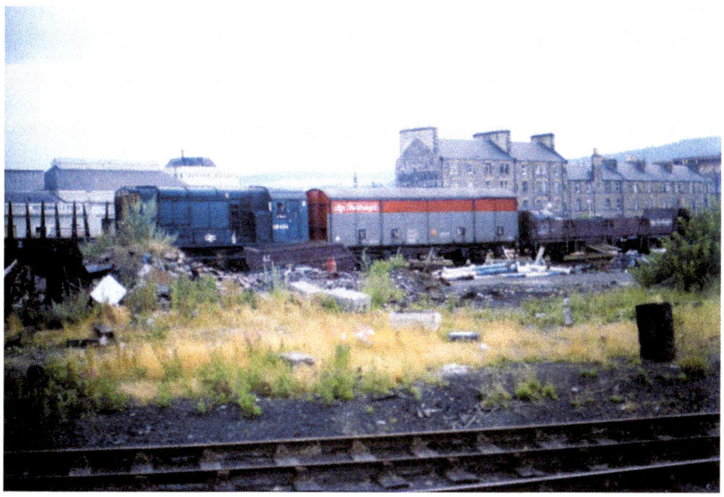

No. 08424 (TJ) at Perth Mineral Yard with an assortment of wagons. Taken August 1984.

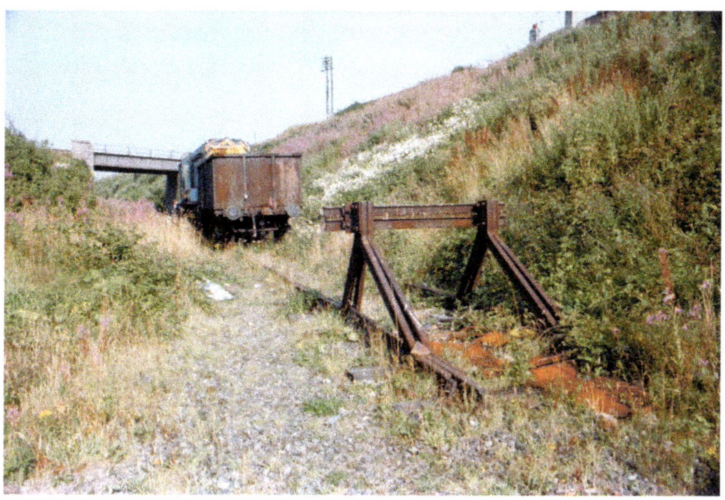

No. 08446 (AY) with a mineral wagon on the stump of the ex-Caledonian route from Stevenston Moor Park to Ardrossan Montgomerie Pier line. The bridge in front of the loco carries the present Ardrossan South Beach to Largs line, which is still currently in use but singled for passenger use. Taken August 1983.

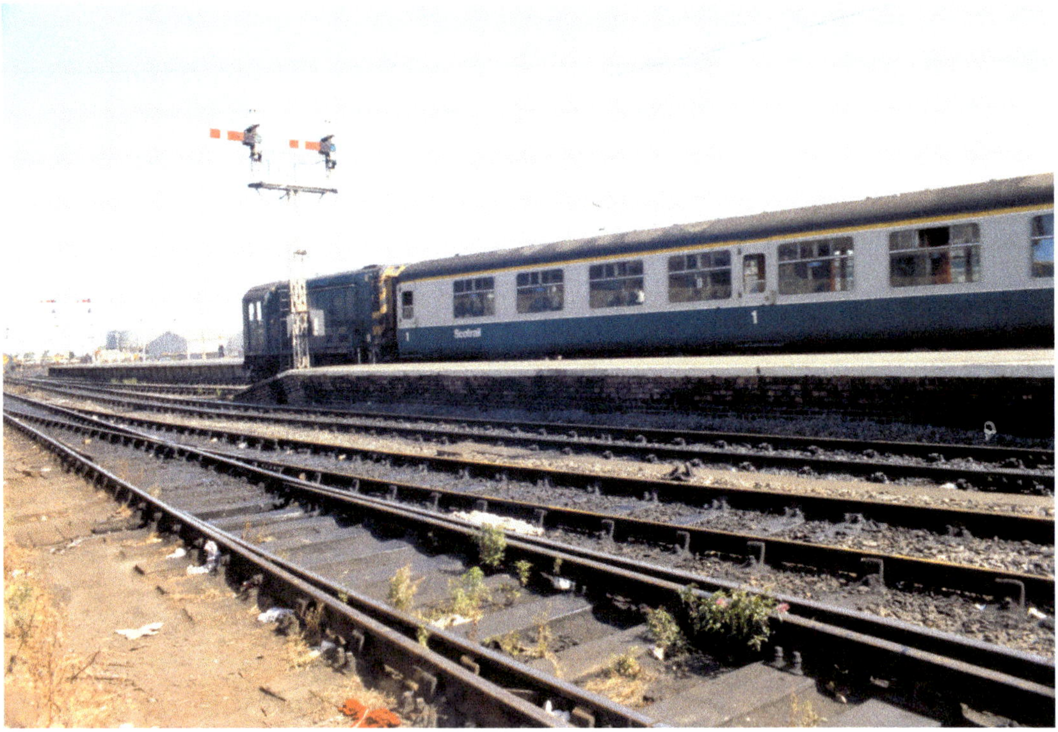

No. 08717 (IS) at Inverness station attaching to a Mk 2 first-class coach. Taken April 1985.

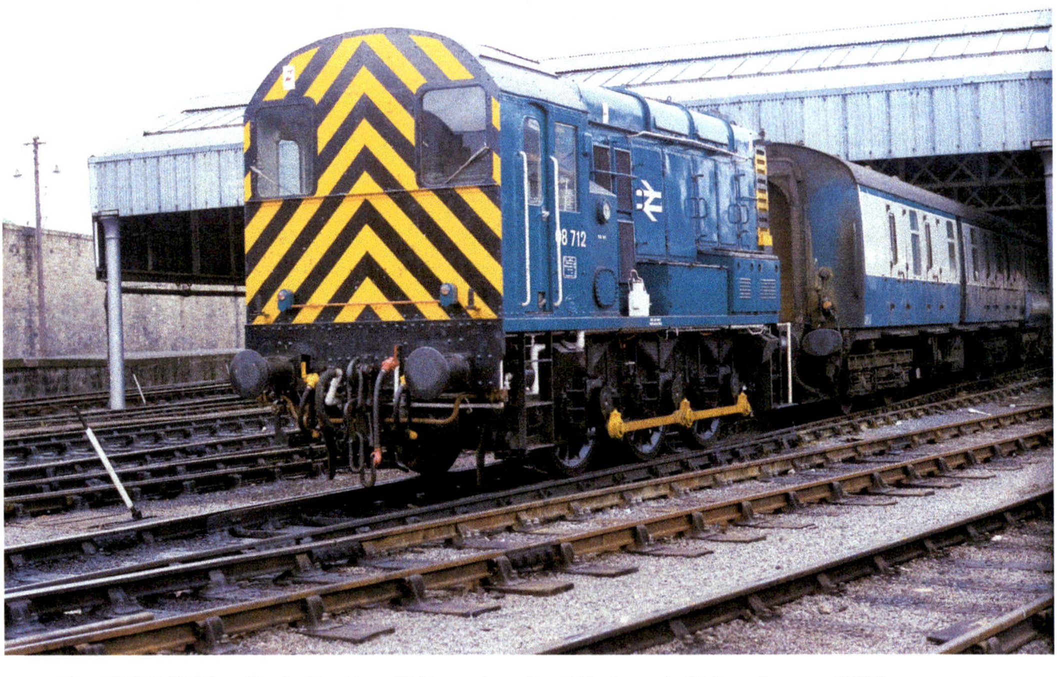

No. 08712 (DE) at Perth Carriage Sidings shunting Mk 1 stock. Taken August 1984.

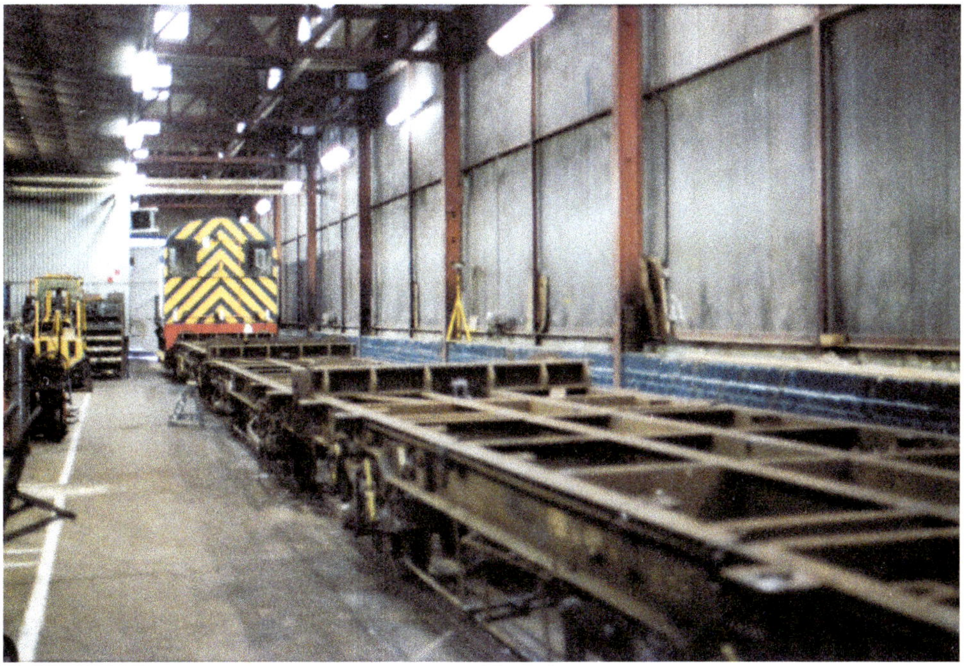

No. 08828 (AB) inside Ferryhill Depot, Aberdeen, with an engineers' wagon. Taken February 1985.

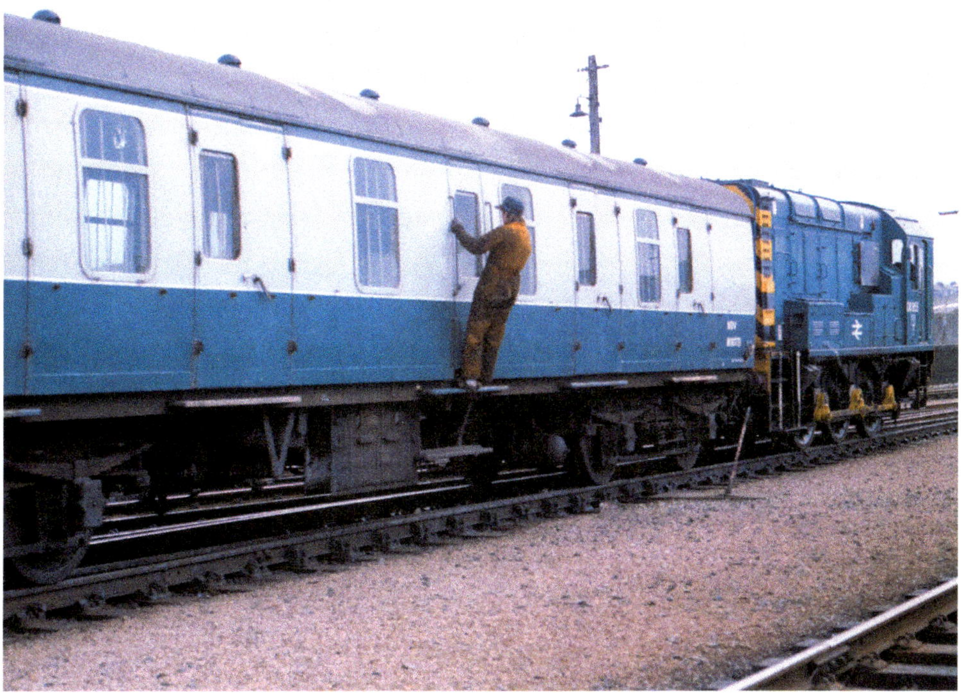

No. 08855 (AB) with a member of the shunting staff at Aberdeen station. Taken February 1985.

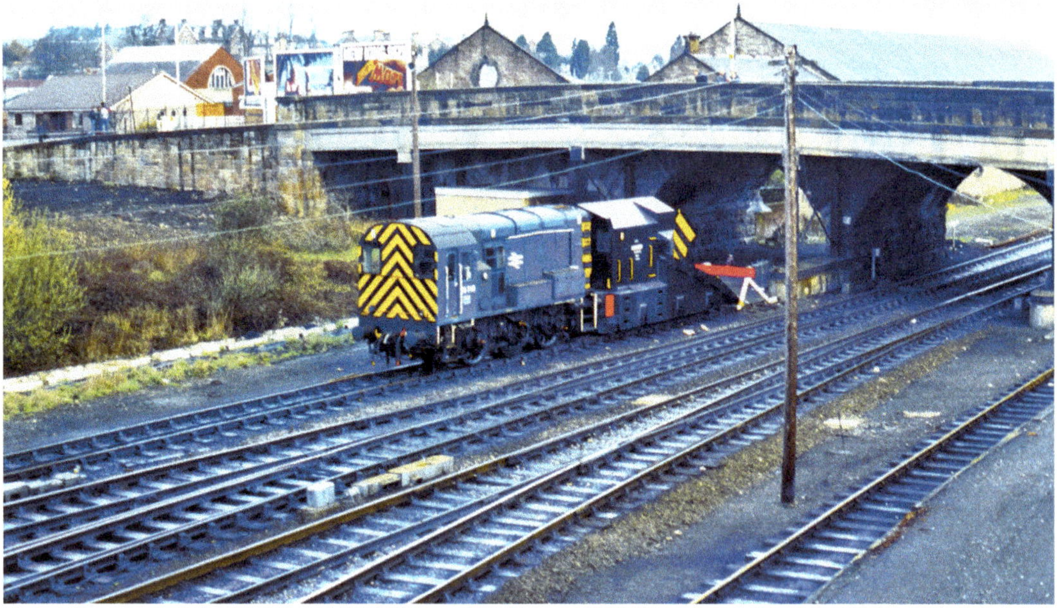

No. 08712 (TJ) stabled at the north end of Perth station with an independent snow plough. Taken April 1985.

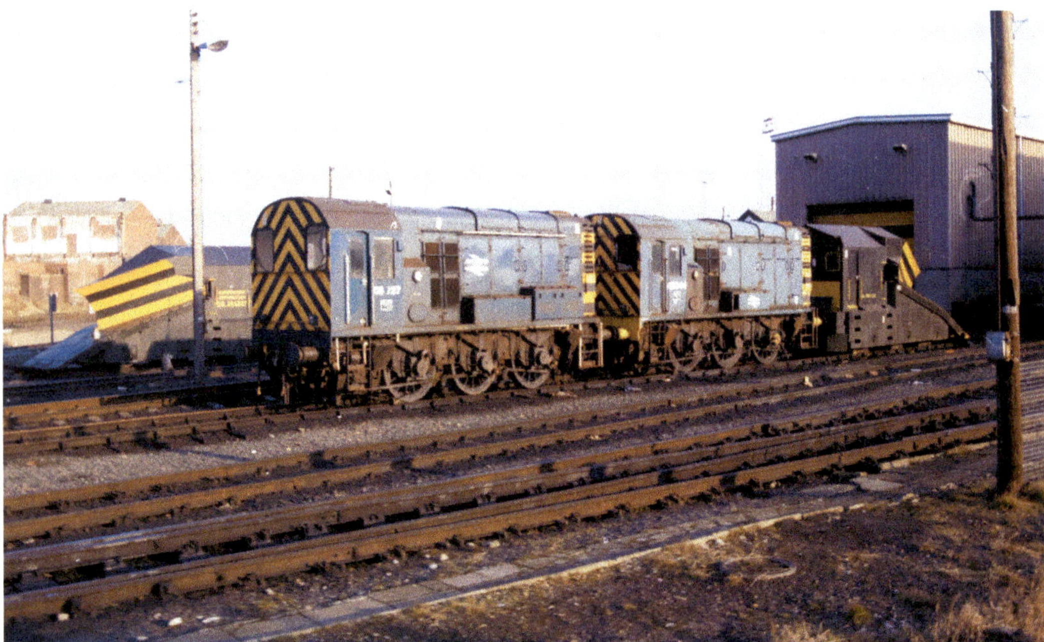

Nos 08727 and 08344 (both AY) with an independent snow plough at Ayr Depot. Taken February 1986.

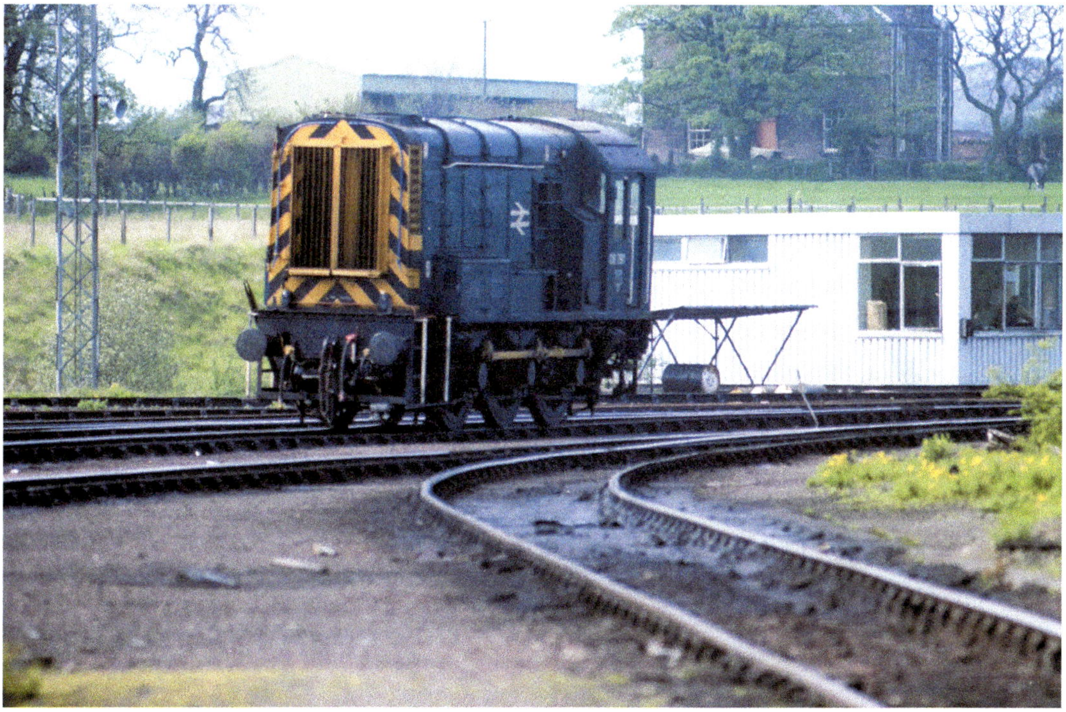

No. 08791 (HA) at Millerhill Yard. It was unofficially named 'South Esk' in March 1985. It was withdrawn from traffic in August 1992 and scrapped at MC Metals, Glasgow, by November 1993. Taken May 1983.

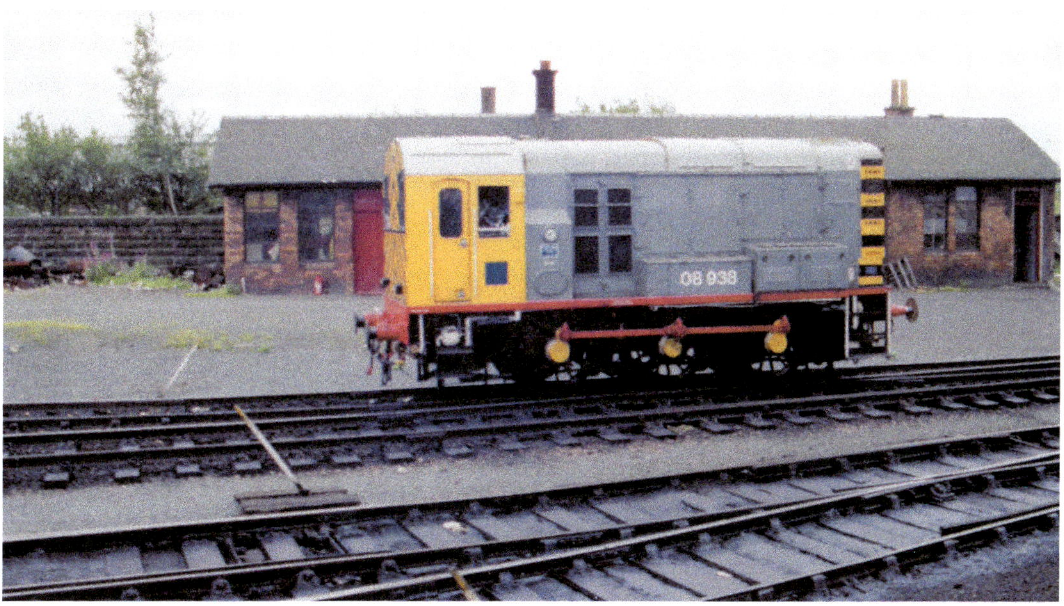

No. 08938 (ML) in Transrail livery stabled at Glasgow's Salkeld Street Parcels Depot. This loco was withdrawn from traffic in December 1995. It was moved by road to Wigan CRDC and broken up by June 2000. Taken August 1986.

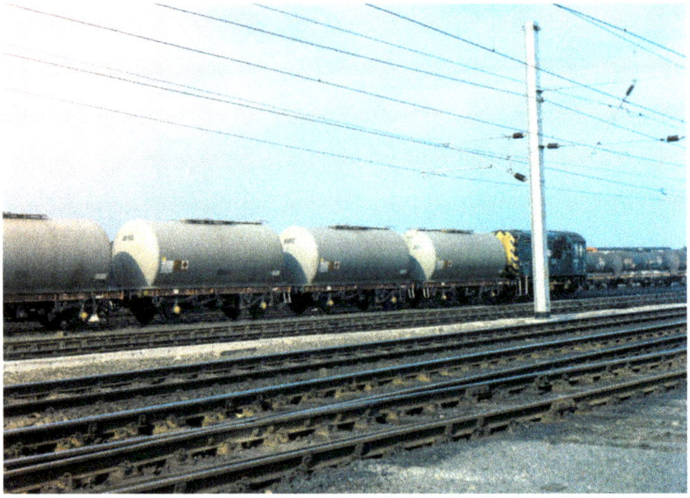

No. 08727 (AY) at Falkland Yard with TTA tanks for Prestwick Airport. Taken March 1987.

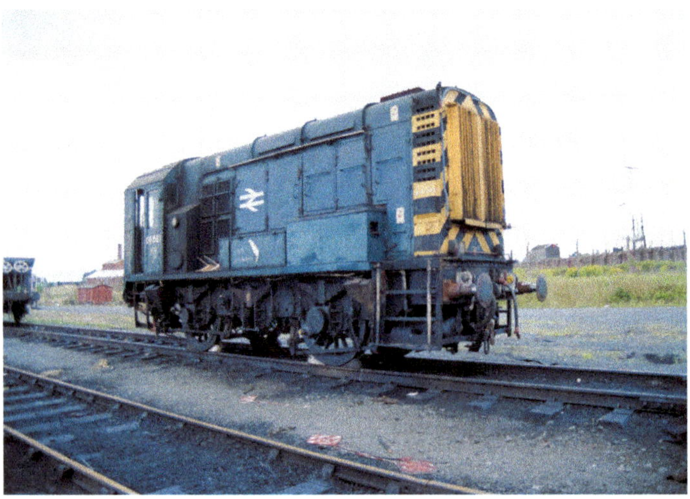

No. 08561 (AY) minus connecting rods at Ayr Depot. It was later transferred to Motherwell Depot in July 1995 but was withdrawn from traffic in May 1998. However, it was reinstated to Crewe Diesel Depot in March 1999. Taken August 1987.

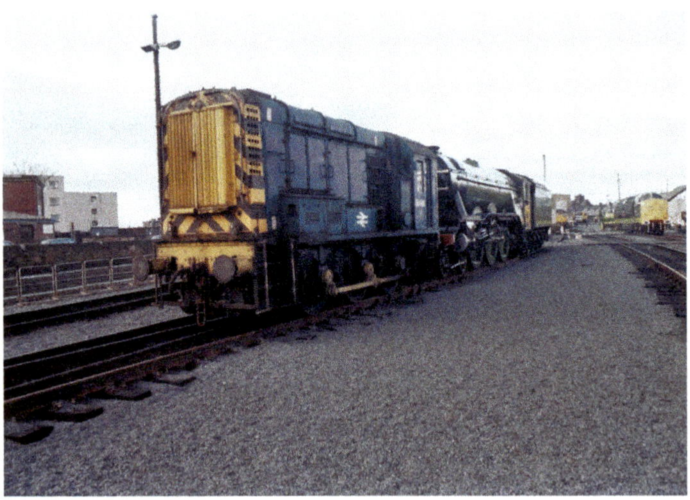

No. 08446 (AY) at Ayr Depot shunting steam loco A3 No. 60103 *Flying Scotsman* for a BR open day. Taken October 1983.

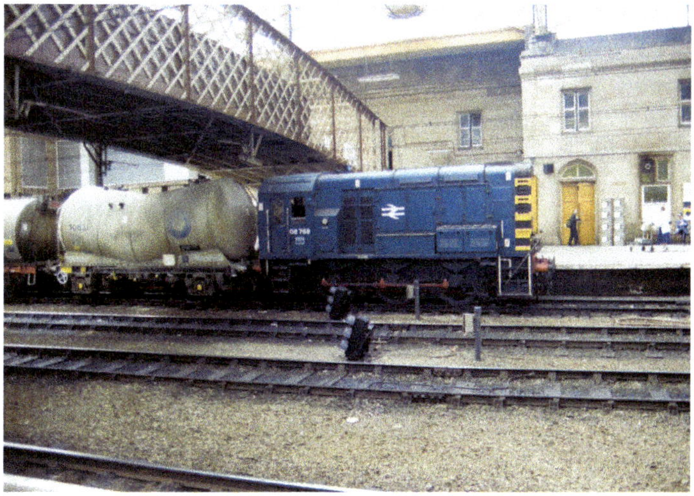

No. 08768 (ML) at Perth station heading north to Perth Yard with cement tanks. Taken June 1988.

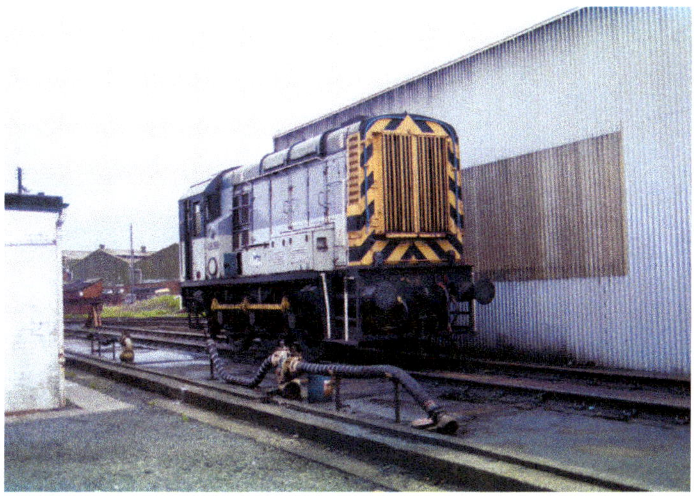

No. 08586 (AY) at the fuelling point at Ayr Depot. Taken September 1993.

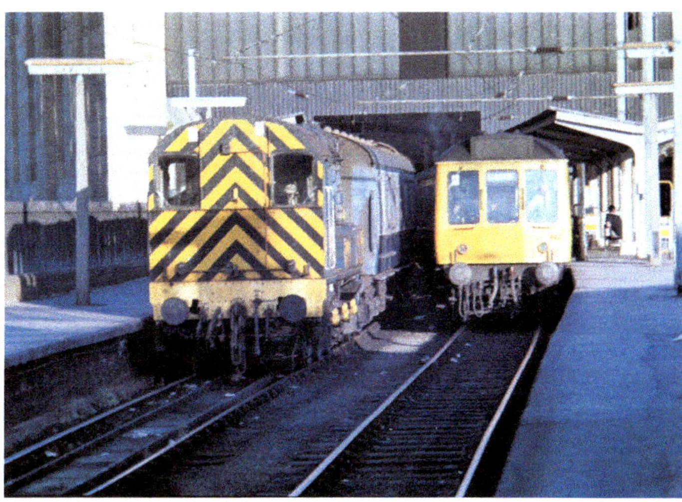

No. 08910 (KD) and three-car No. 107447 (AY) at Platforms 8 and 7 at the north end of Carlisle station. Taken August 1981.

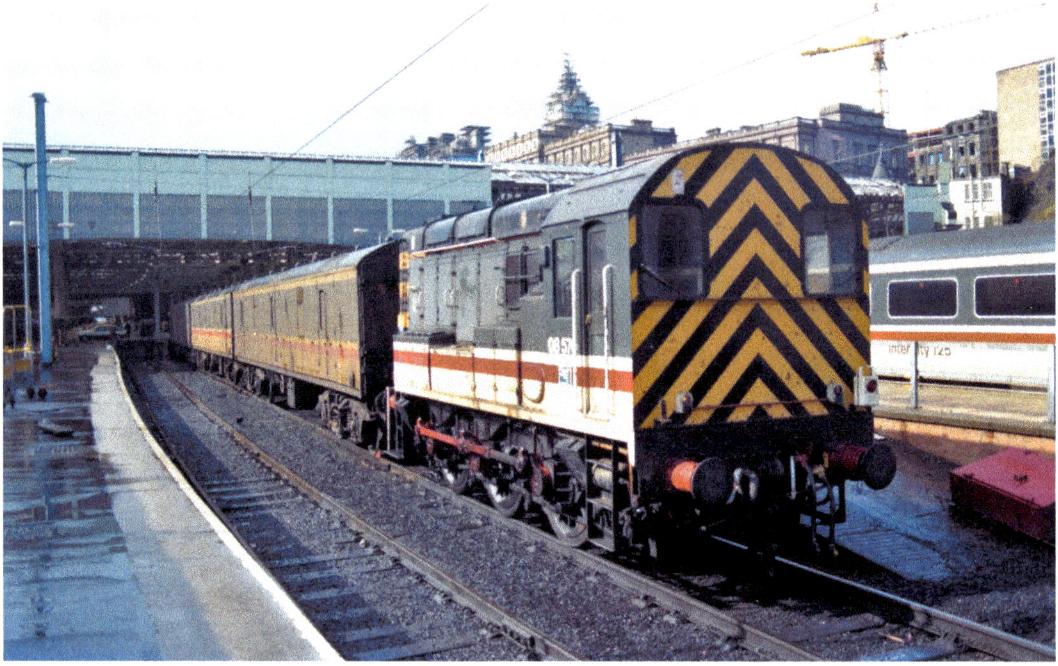

No. 08576 (HA) at Edinburgh with Motorail vans. Taken March 1990.

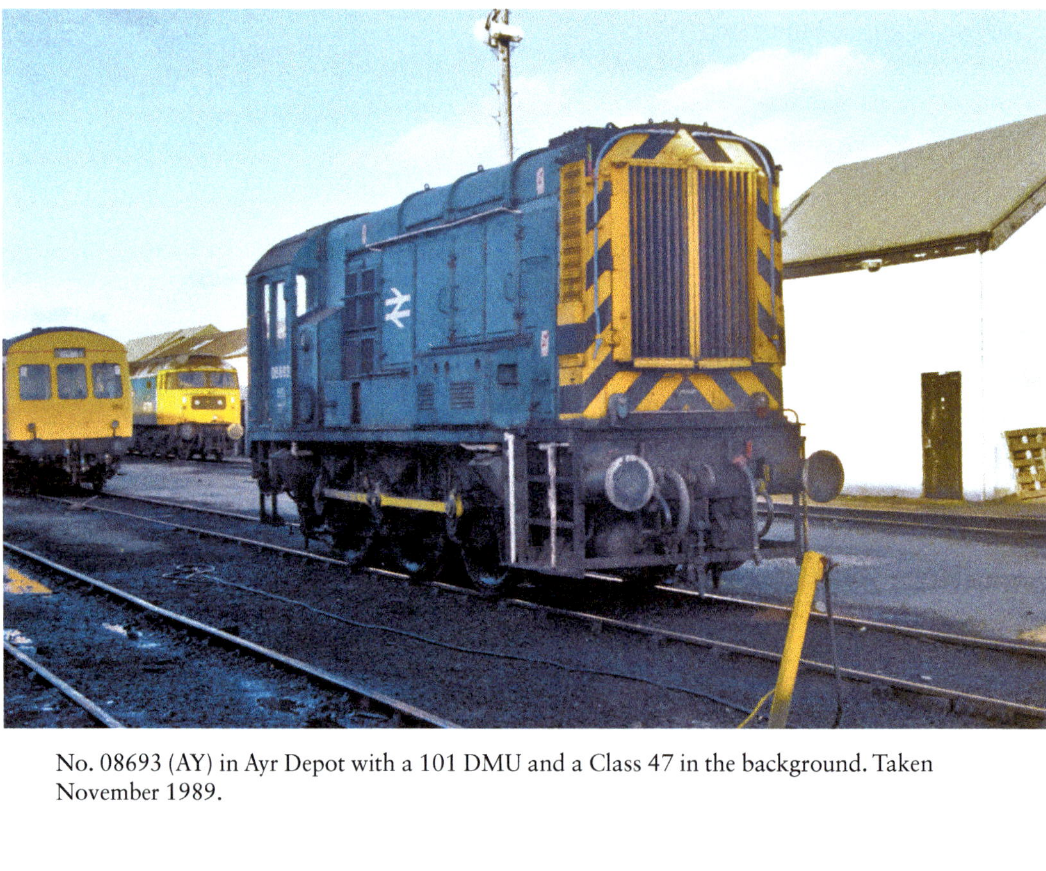

No. 08693 (AY) in Ayr Depot with a 101 DMU and a Class 47 in the background. Taken November 1989.

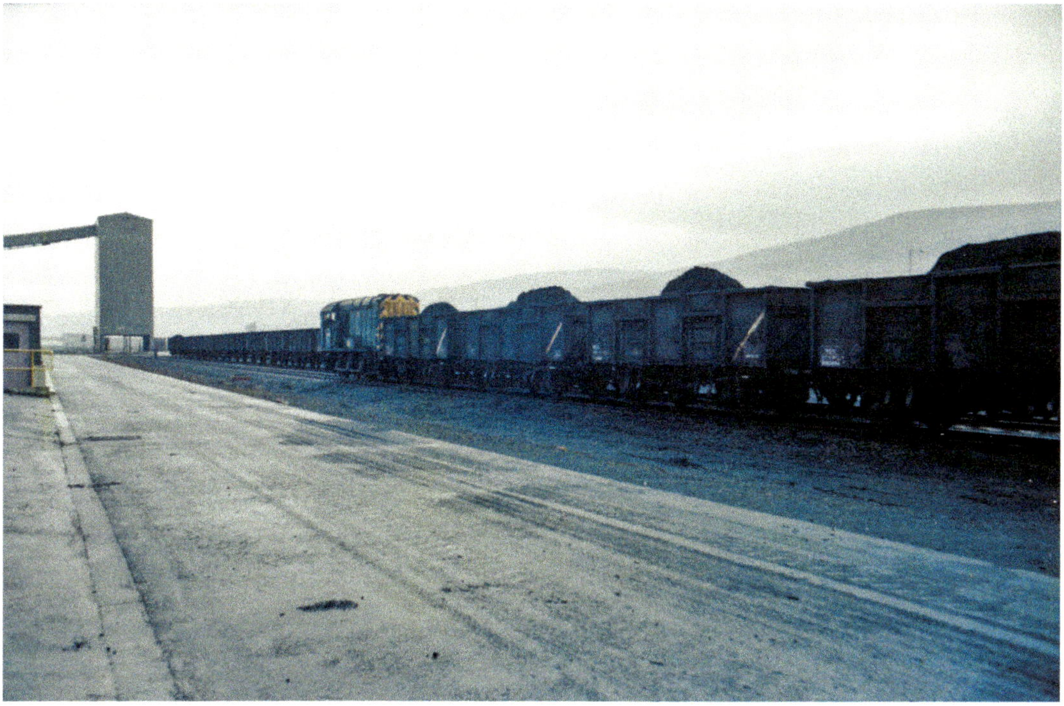

No. 08449 (AY) at Hunterston Low Level shunting loaded mineral wagons. Taken December 1983.

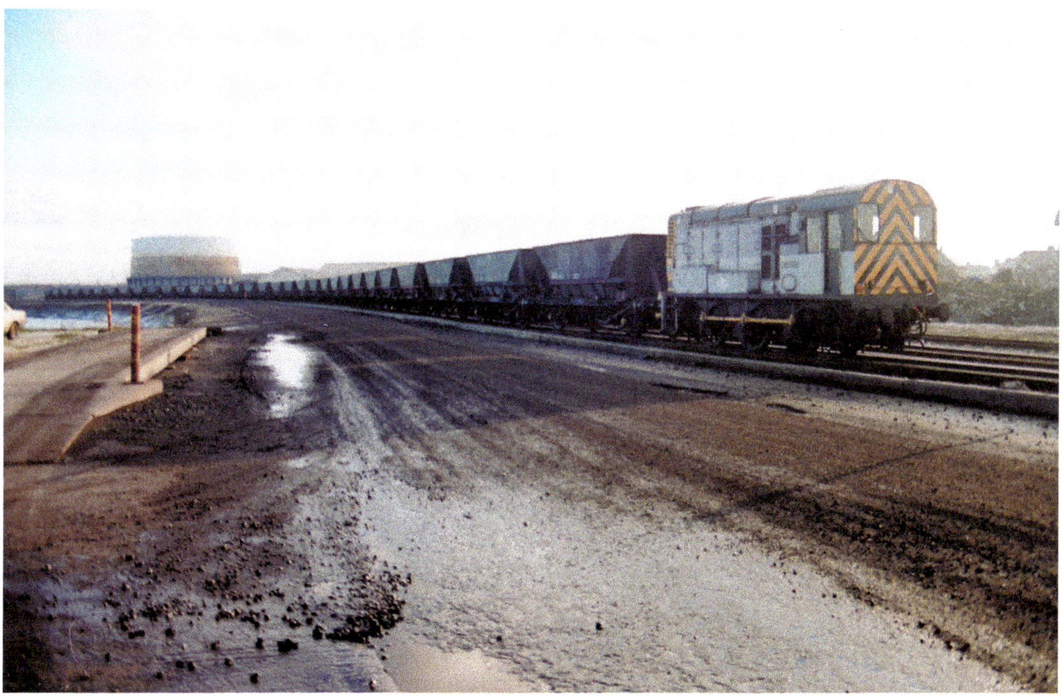

No. 08586 (AY) at Ayr Harbour with MGR empties. Taken October 1990.

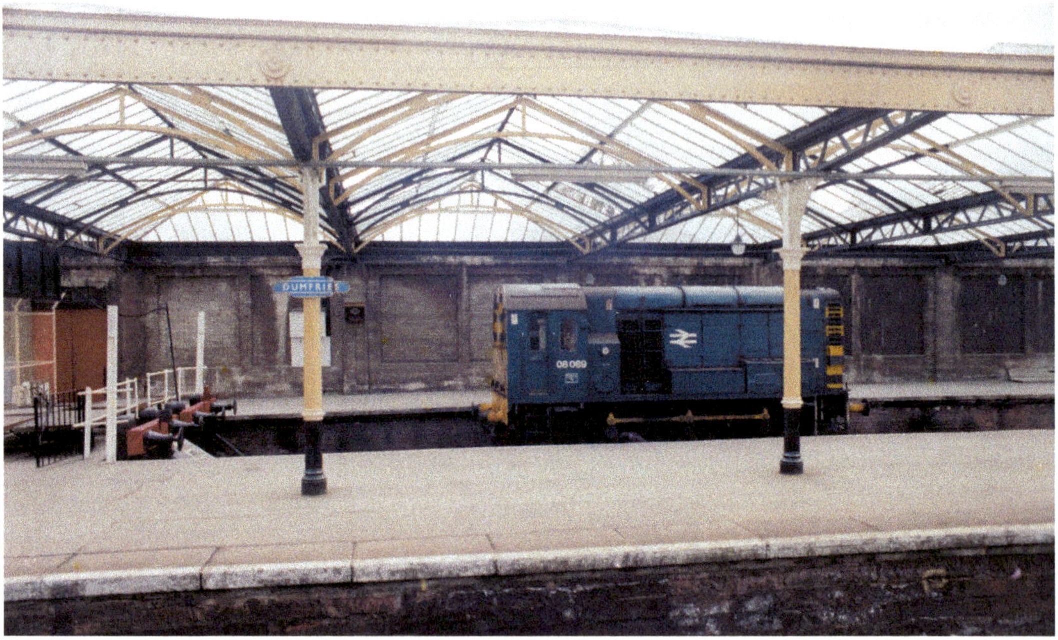

No. 08069 (KD) in the bay platforms at Dumfries. The Carlisle Kingmoor-based 08s were outbased regularly at Dumfries to shunt the local yard. These platforms were removed in the early 1990s. Taken May 1983.

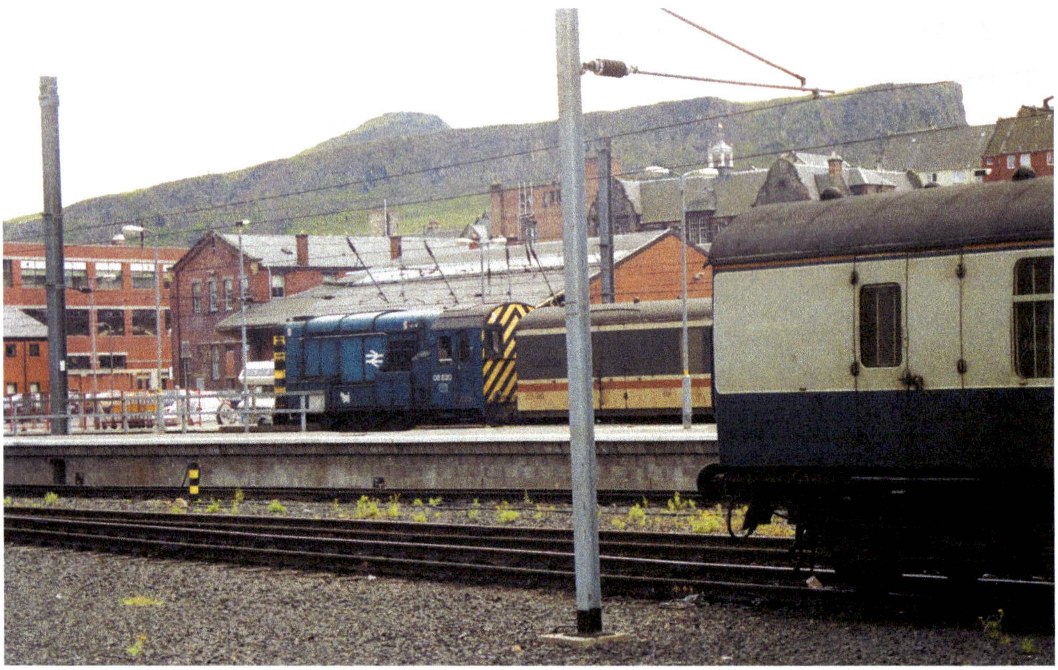

No. 08620 (HA) at Edinburgh with a Motorail van. Note Arthur's Seat in the background. This is part of an extinct volcano. Some people believe its name was derived from King Arthur. Taken September 1991.

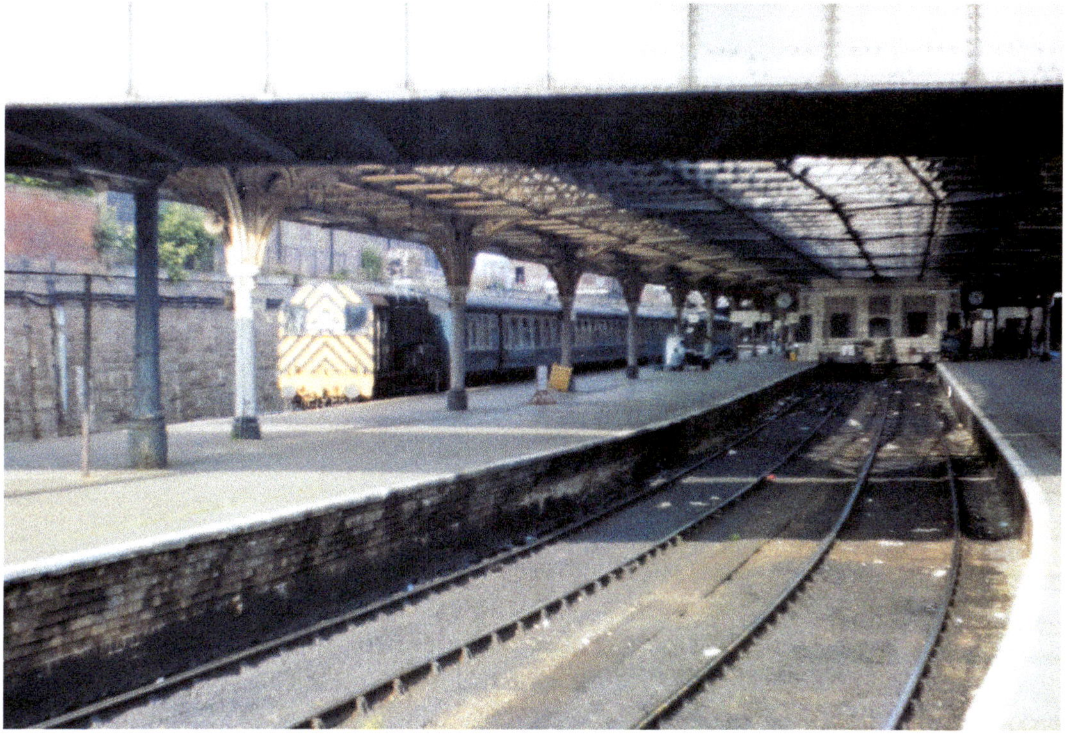

No. 08761 (DE) shunting passenger coaches at Dundee station. Taken August 1981.

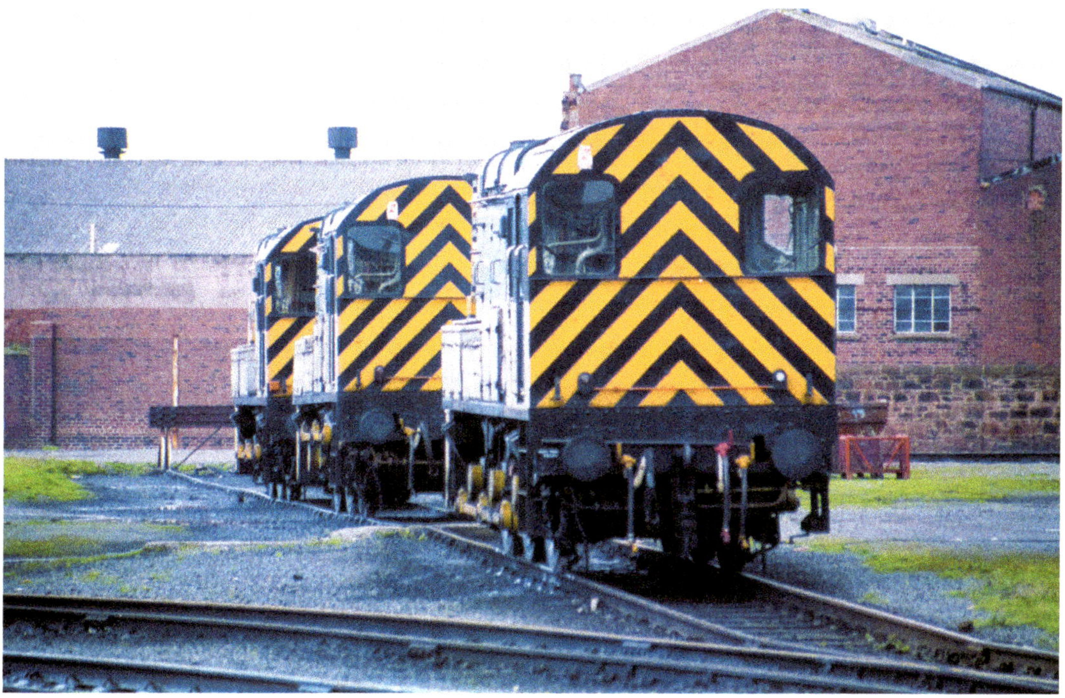

Three Class 08s at Ayr Depot stabled next to the fuelling road. Taken April 1992.

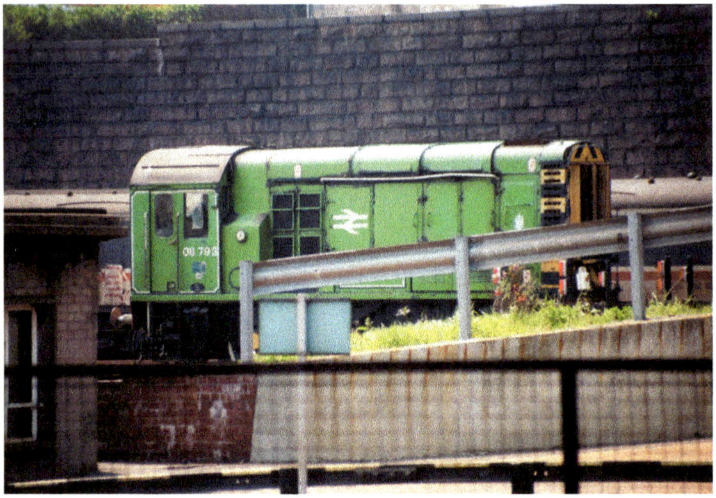

No. 08793 (AB) at Clayhills Sidings, Aberdeen. Taken May 1992.

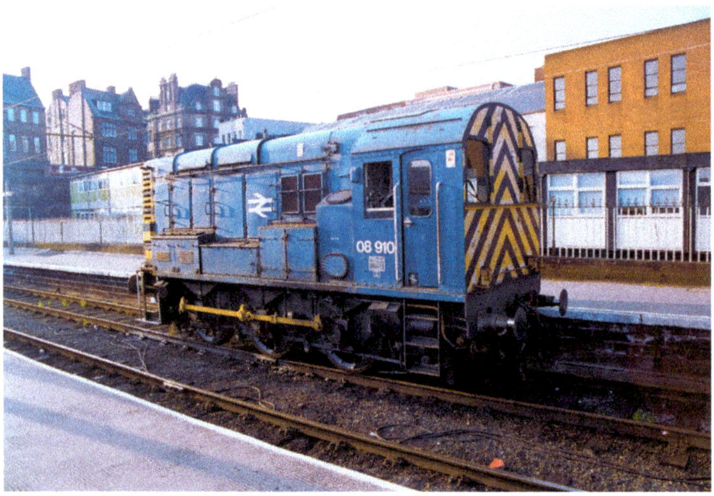

No. 08910 (KD) at the north end of Carlisle station at Platform 8 between shunting duties. Taken May 1992.

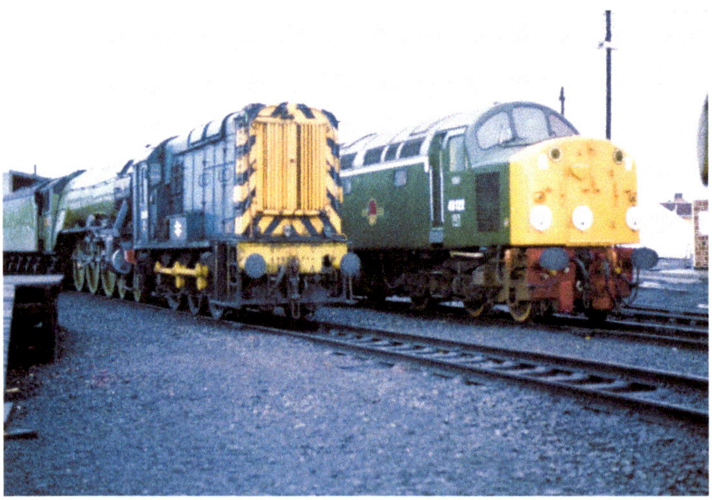

Nos 08345 (AY) with 40122 (KD) and A3 steam loco No. 60103 *Flying Scotsman* at Ayr Depot for a BR open day. Taken October 1983.

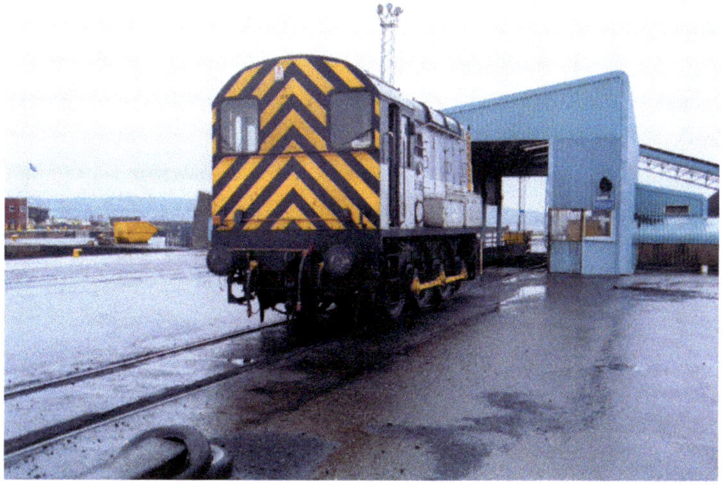

No. 08586 (AY) at Ayr Harbour standing near the wagon weighbridge. Taken May 1993.

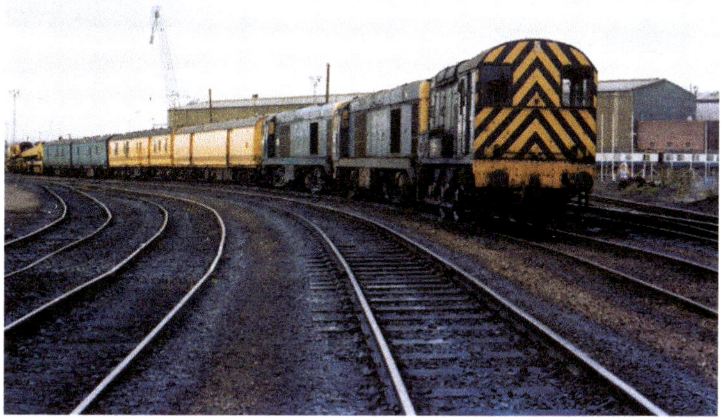

A Class 08 with Class 20s and engineers' breakdown coaches at Motherwell Depot. Taken April 1982.

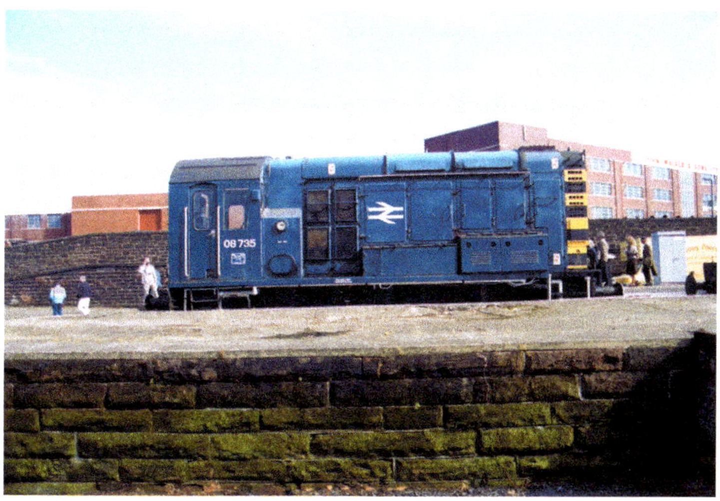

No. 08735 (AY) at Hunslet Barclay Works, Kilmarnock, for a ScotRail open day. Taken September 1988.

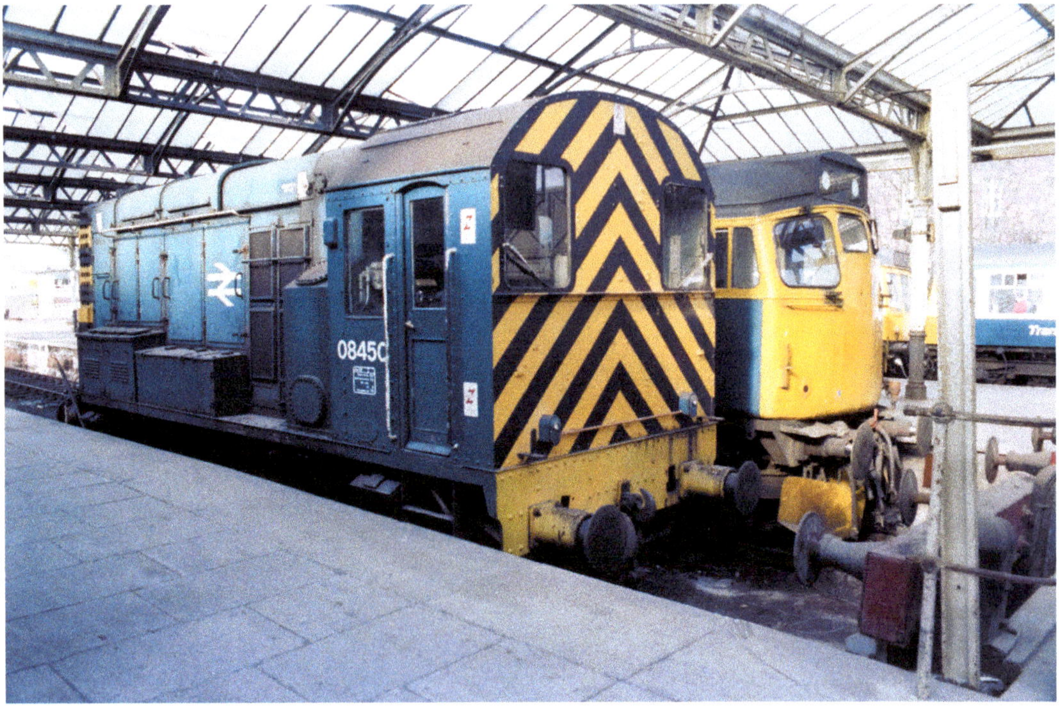

No. 08450 (KD) stabled at the north-end bay platform at Dumfries station in-between shunting duties with No. 27106 (ED). These platforms were infilled in the early 1990s. Taken April 1982.

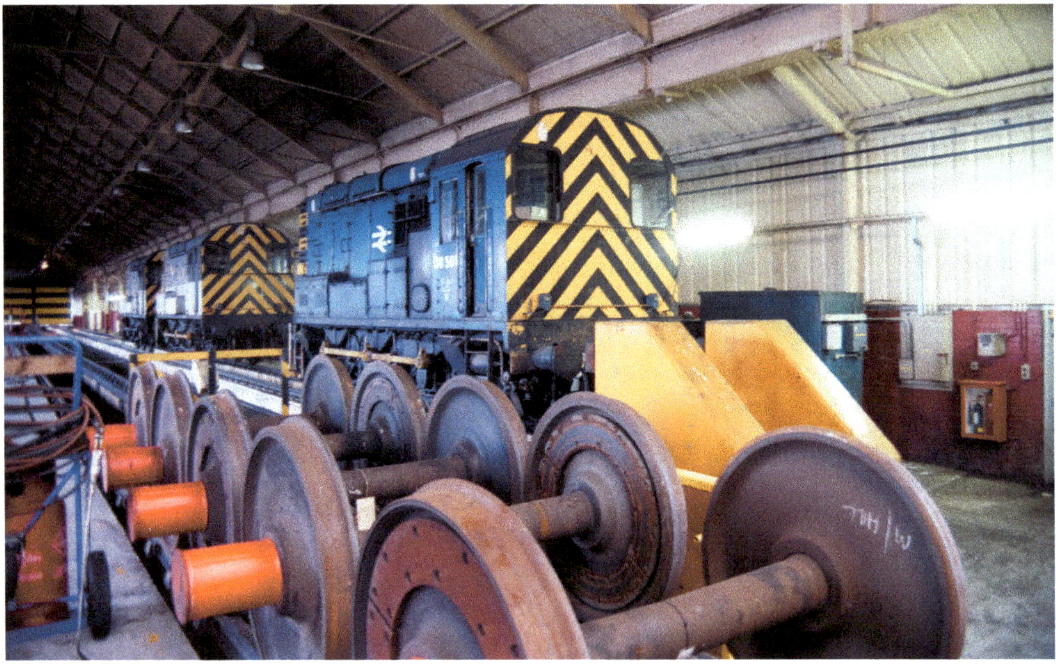

No. 08561 (HQ) and two other classmates inside Ayr Depot. This depot was closed by DB Schenker in 2014 and finally demolished in 2019. Taken April 1995.

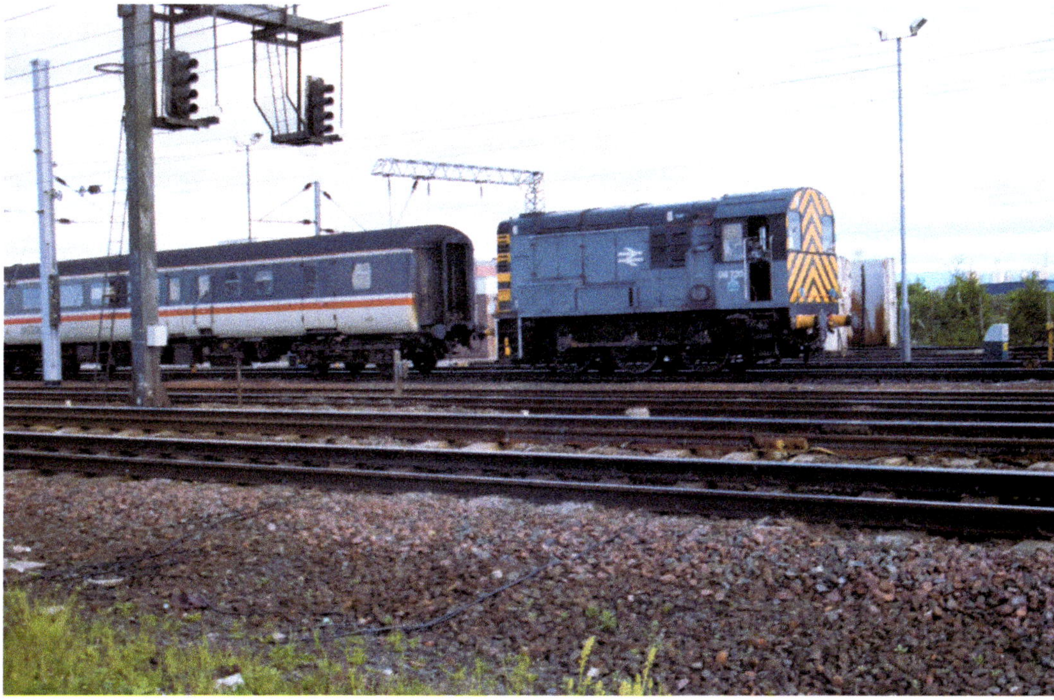

No. 08735 (ED) at Polmadie Depot, Glasgow, shunting InterCity Mk 2 coaches. Taken April 1995.

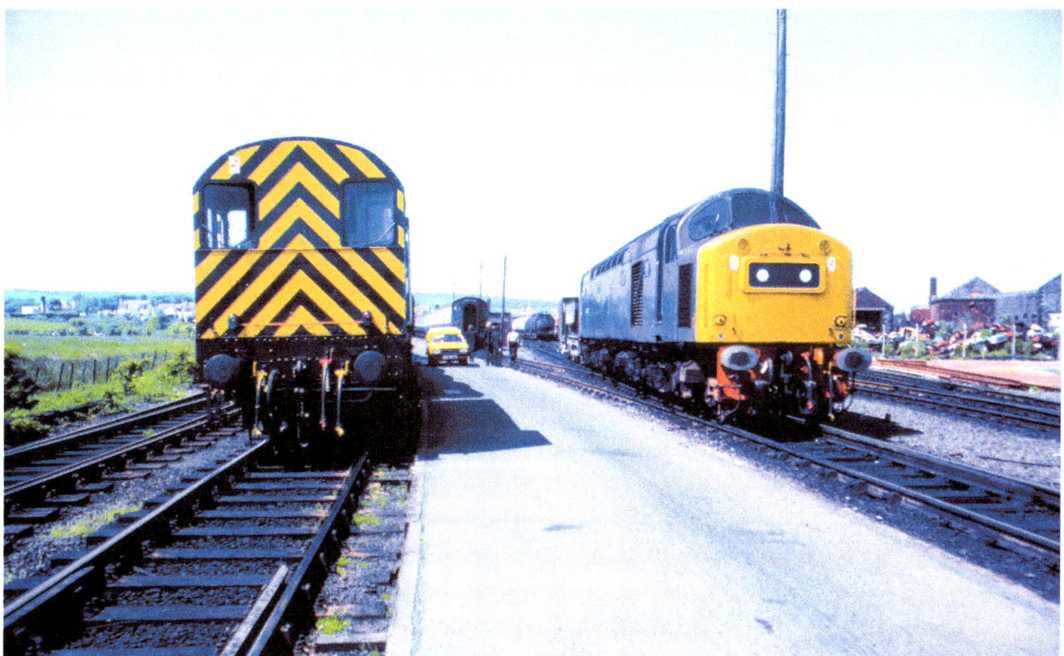

Nos 08733 (AY) and 40177 (CD) at Stranraer Town. The 08 had just shunted in the overnight stock from London Euston and the Class 40 was about to depart with a northbound engineers' train. Taken June 1984.

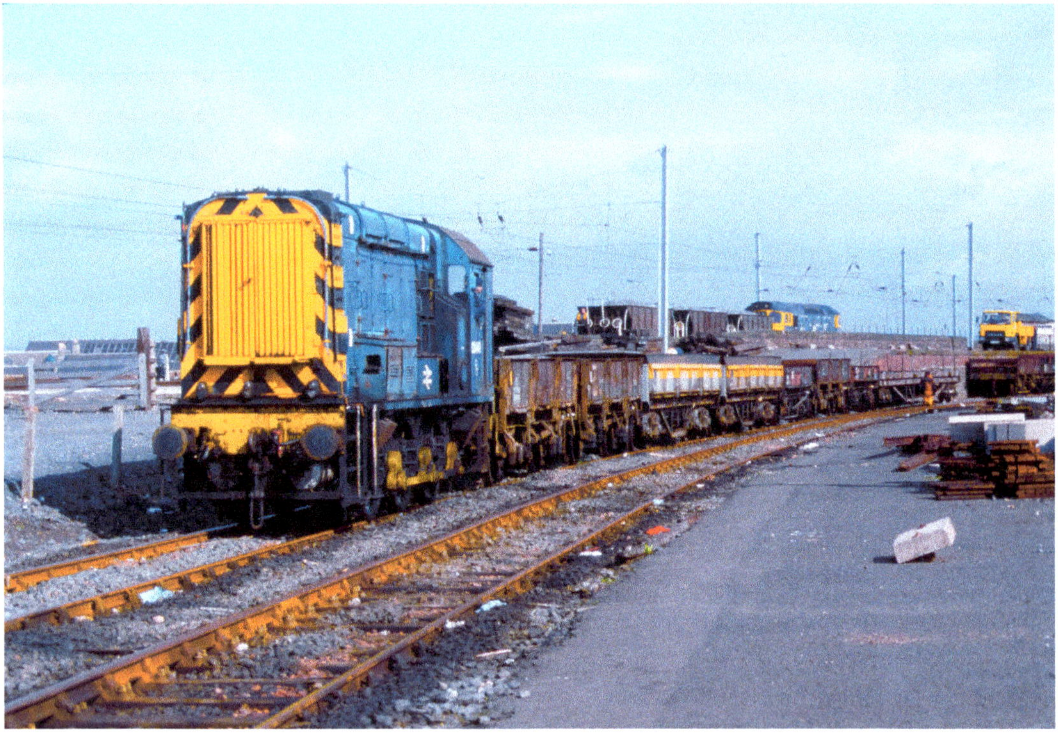

No. 08448 (AY) shunting engineers' wagons at Irvine East Yard. Taken April 1986.

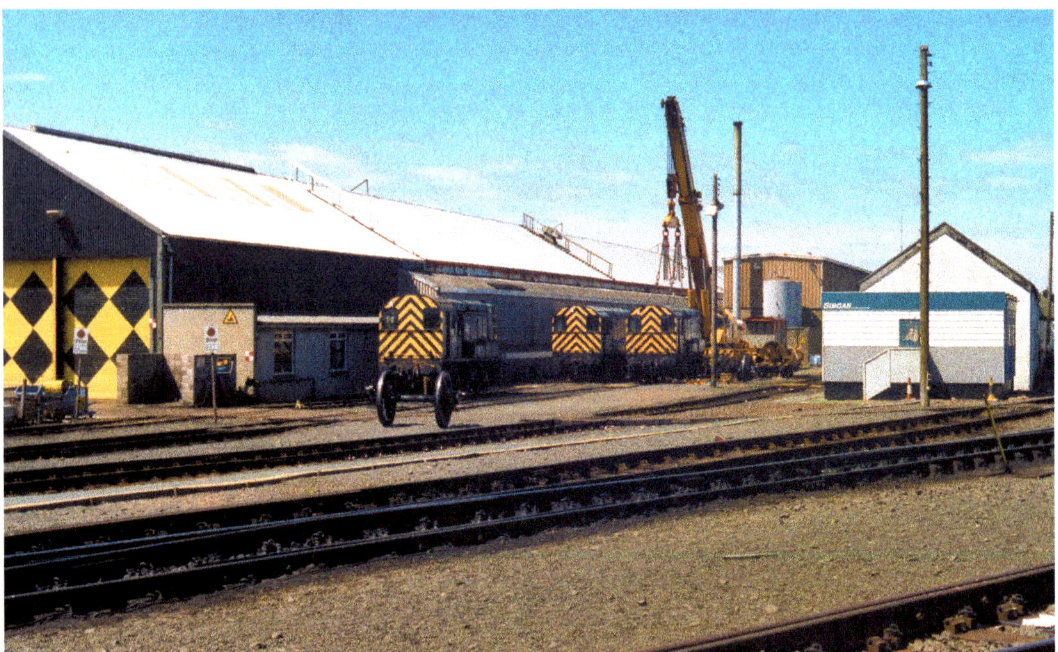

Three Class 08s at Ayr Depot plus a lifting crane in use for wheelset renewal to one of the aforementioned 08s. Taken July 1995.

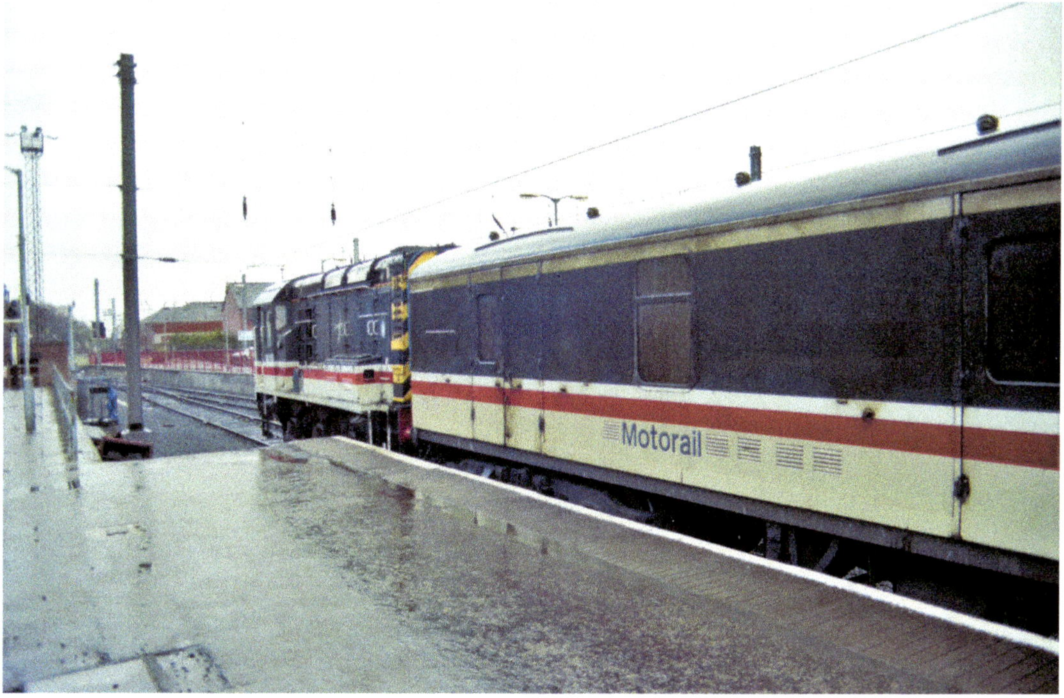

No. 08570 (HA) with a Motorail van at Edinburgh. Taken March 1990.

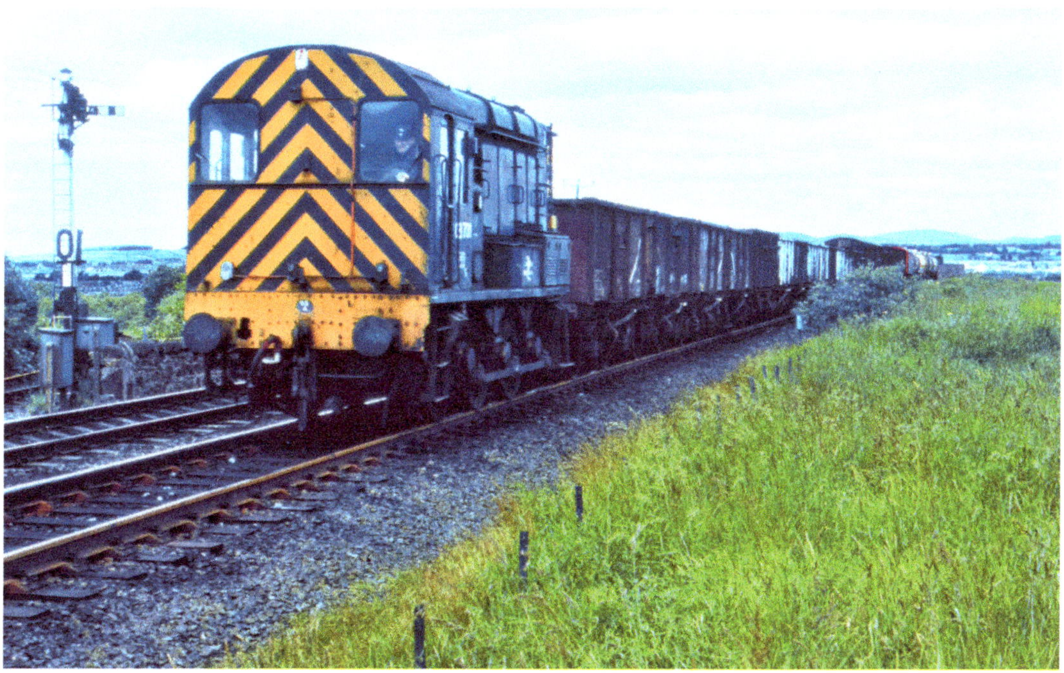

No. 08731 (AY) approaching Bogside, north of Irvine, with a local trip working heading to Falkland Yard, Ayr, where all the wagons would be shunted onto various forwarding services to the south. Taken September 1980.

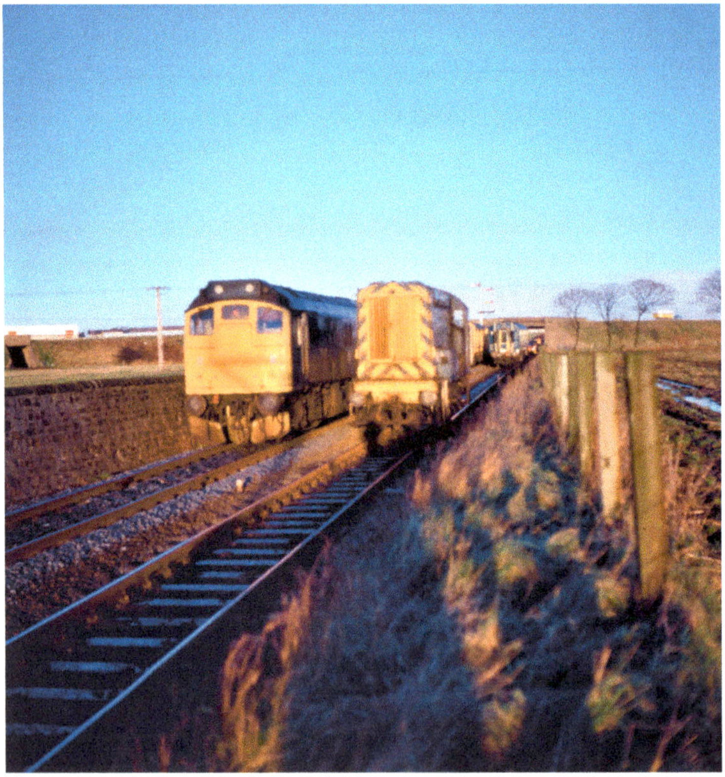

Left: A Class 08 and 25 at Dubbs Junction south of Kilwinning on the Largs line. Both locos had been summoned with the breakdown cranes to rerail Class 107 power car SC 52022, which had been derailed earlier. Taken November 1980.

Below: Class 08 plus Class 126 DMUs at Ayr Depot. Taken September 1981.

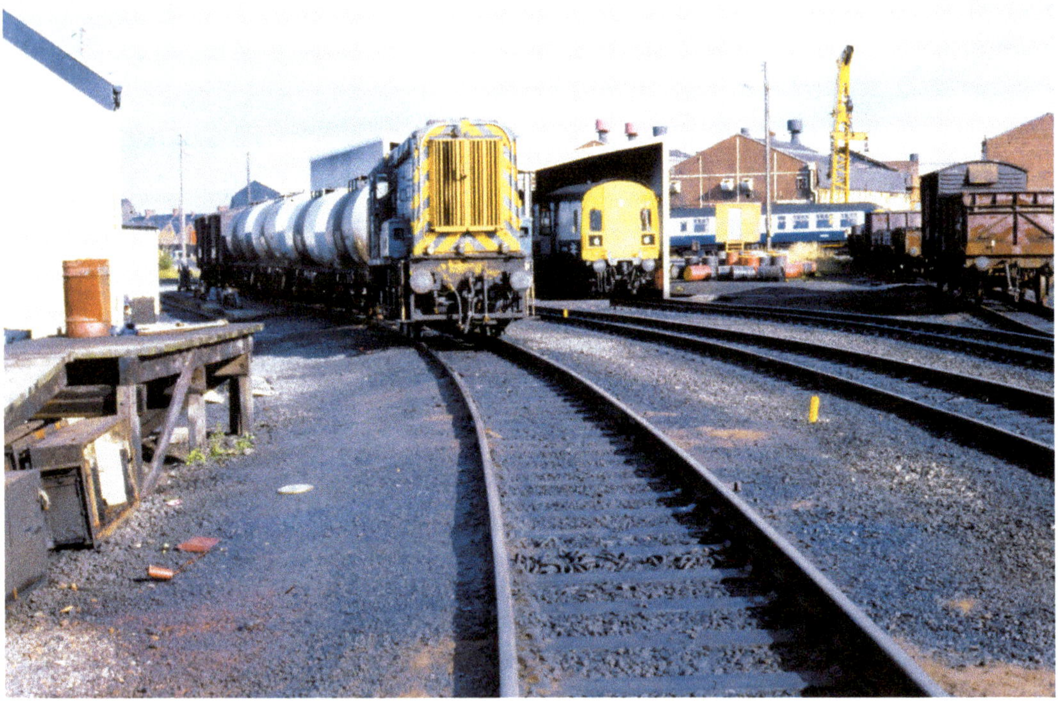

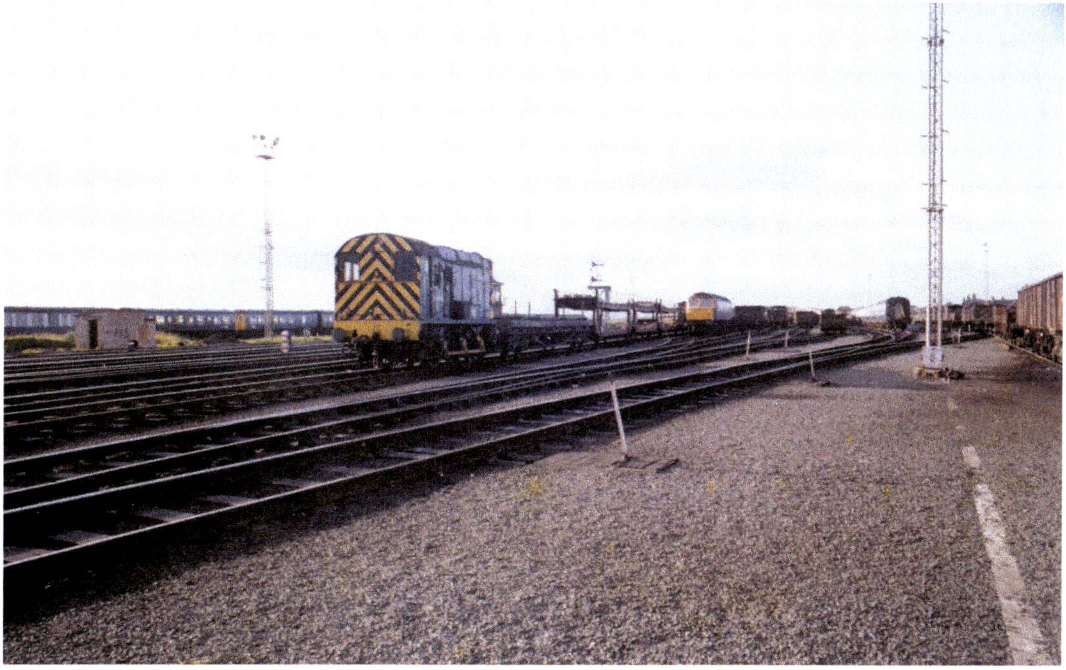

No. 08345 (AY) at Falkland Yard shunting Cartic wagons, plus a Class 47 and withdrawn DMUs. Taken June 1983.

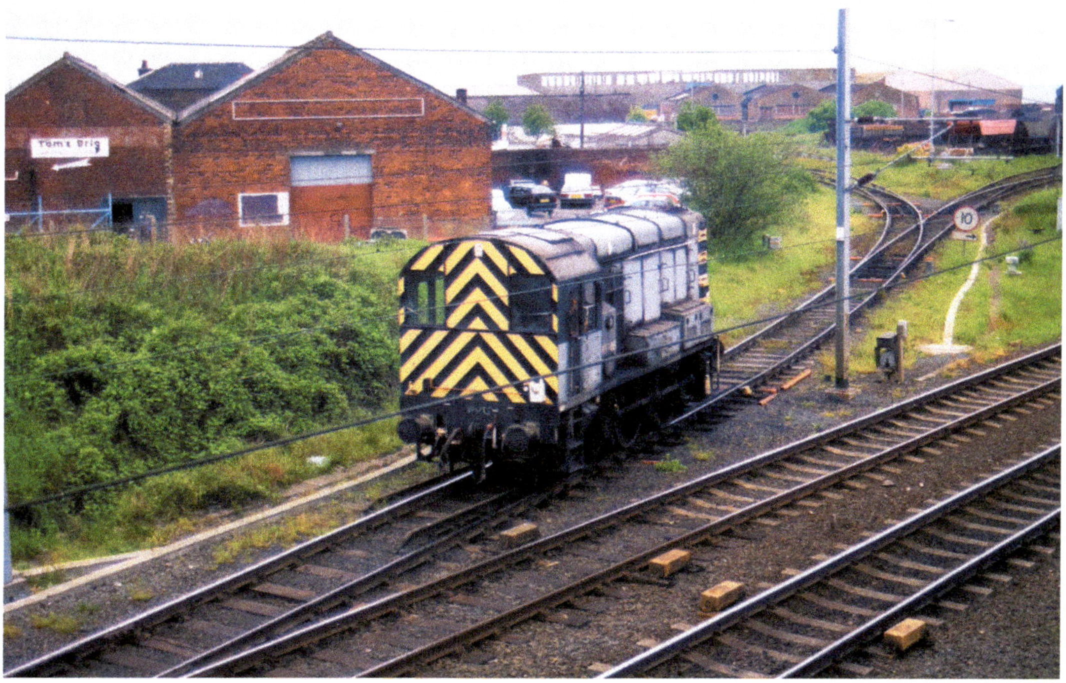

No. 08746 (ML) at Newton-on-Ayr running light from Ayr Depot to Ayr Harbour. Taken June 1998.

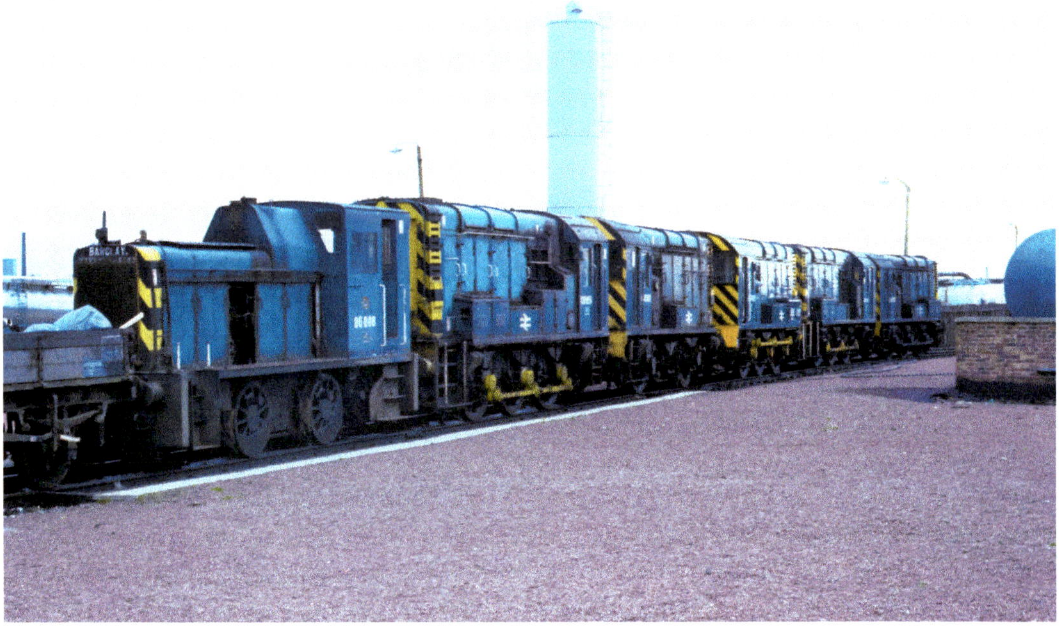

No. 06008 (HQ) plus Nos 08551/621/736/727 (all ML) stabled in-between duties at Polmadie Depot, Glasgow. The 06 was moved shortly afterwards and was scrapped and cut up on-site by H. Campbell of Airdrie. Taken April 1982.

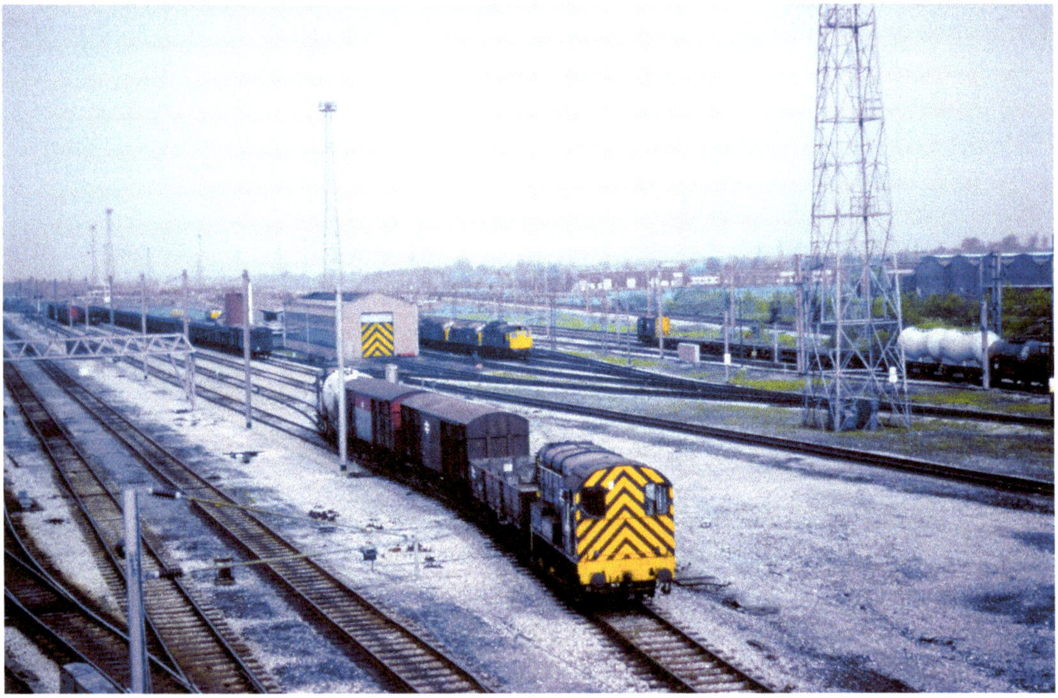

Class 08s in Carlisle's Kingmoor Yard shunting local freight traffic. Class 25, 27 and 40s are also in the background. Taken May 1983.

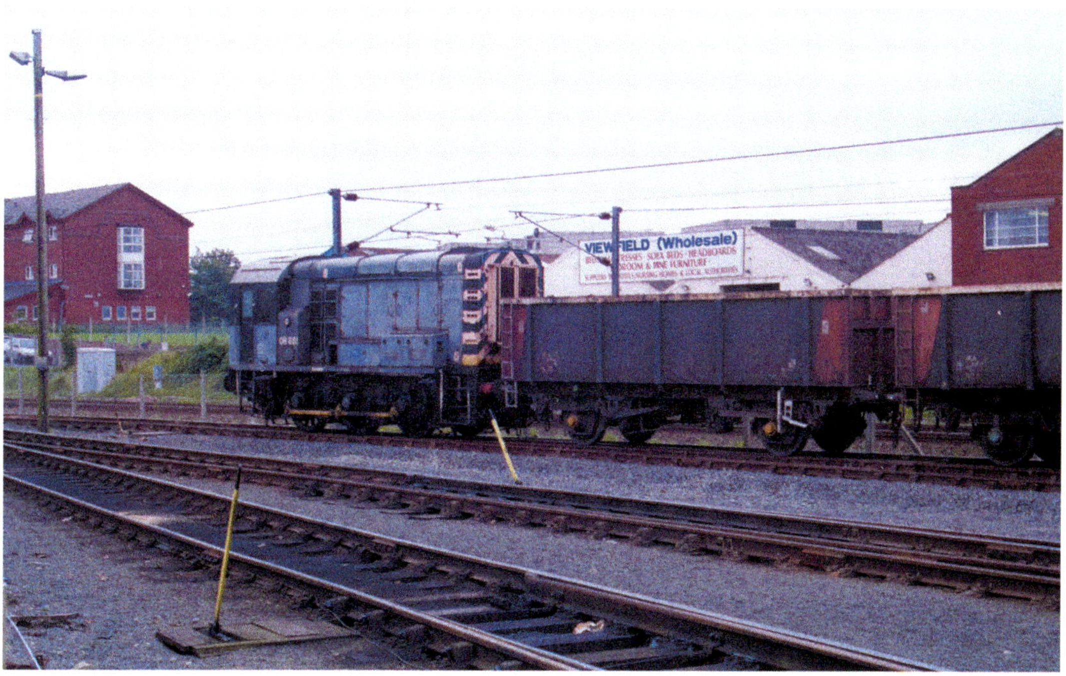

No. 08881 (ML) at Ayr Depot shunting JKA mineral wagons for maintenance. Taken September 1998.

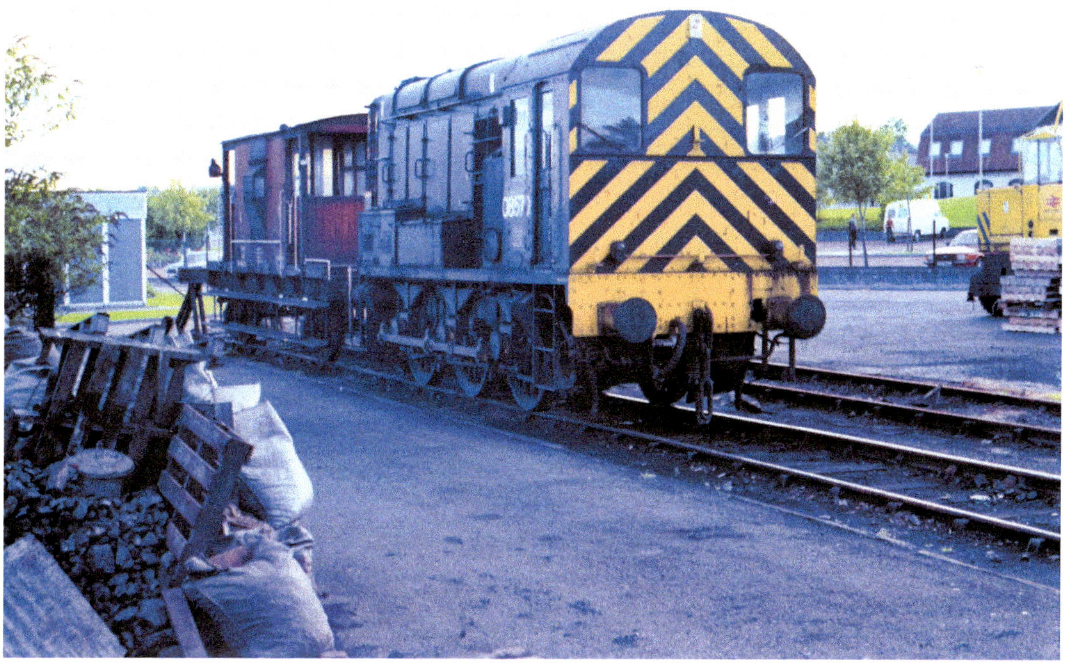

No. 08570 (DE) with a brake van in Kirkcaldy Yard between duties. Taken August 1981.

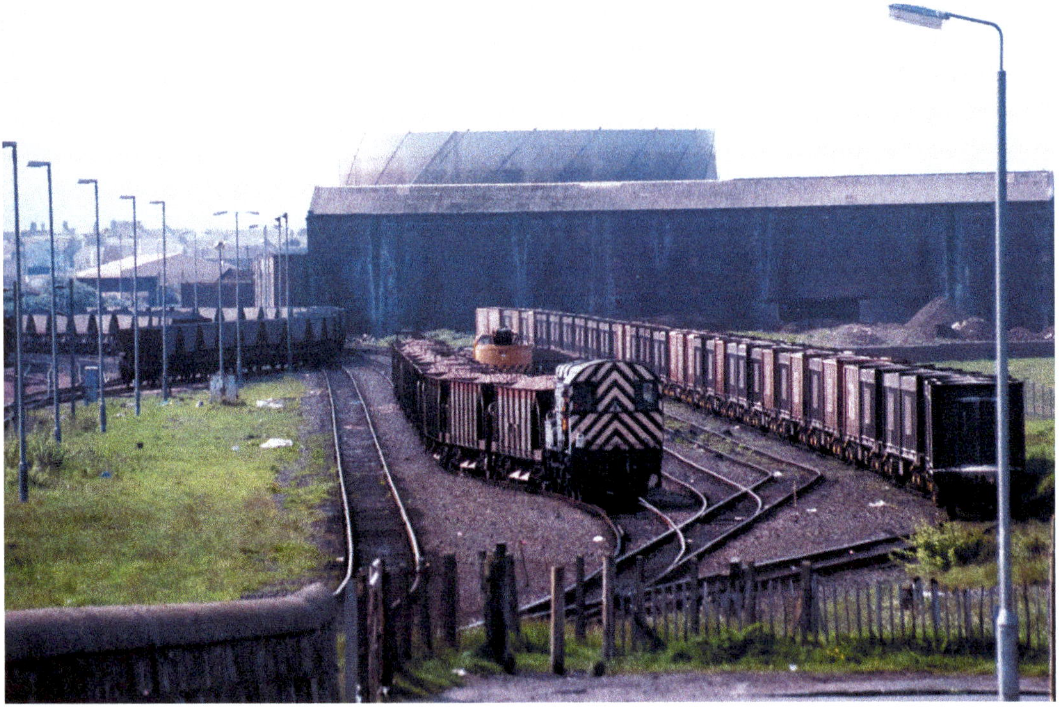

Class 08 with an engineers' train in Falkland Yard. These wagons had just been brought in from Hillhouse Quarry, near Barassie. Hillhouse Quarry provided ballast to the local area for many years, but Network Rail withdrew the contract in 2001. Taken July 1998.

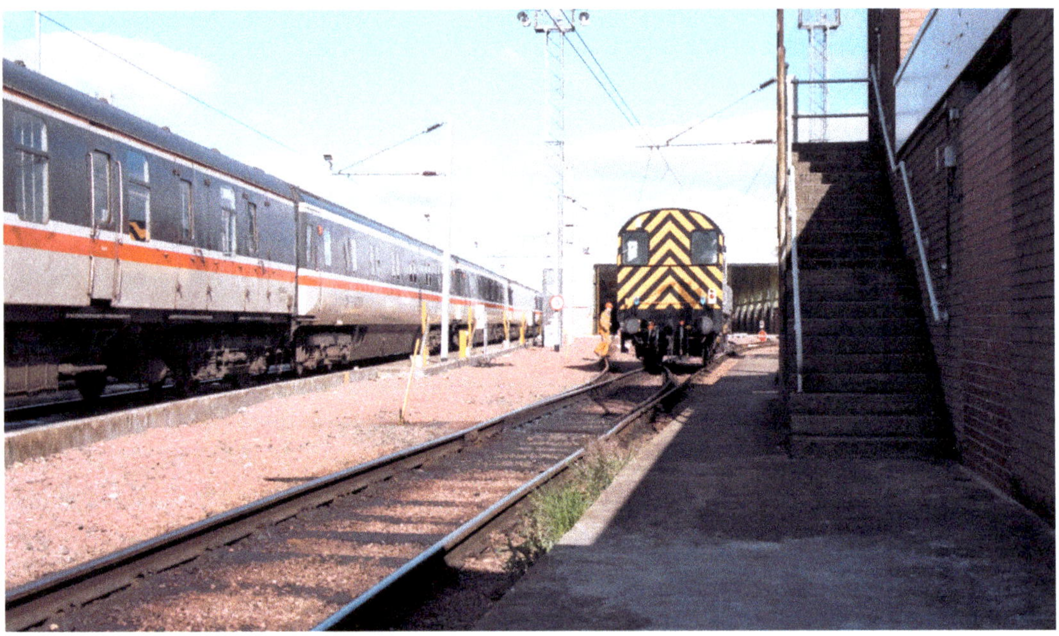

No. 08735 (AY) at Polmadie Depot, Glasgow, with Mk 3 Sleeper stock. This stock was eventually replaced by Mk 5 stock from 2018. Taken April 1995.

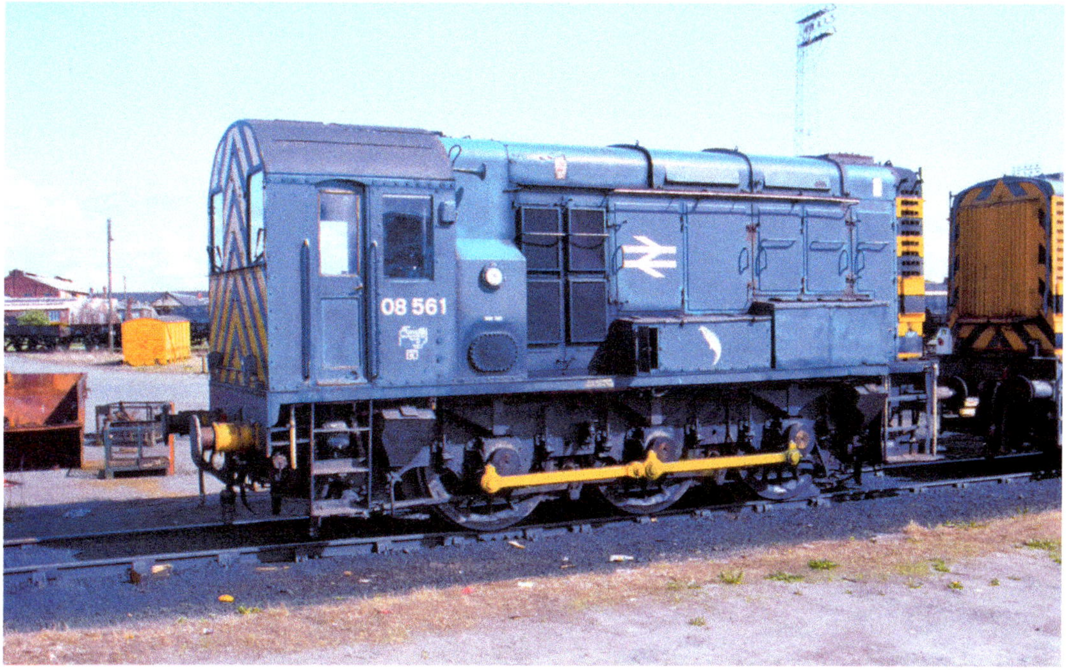

No. 08561 (AY) at Ayr Depot. This loco was in service at Ayr until being transferred to Motherwell Depot in July 1995. Taken May 1993.

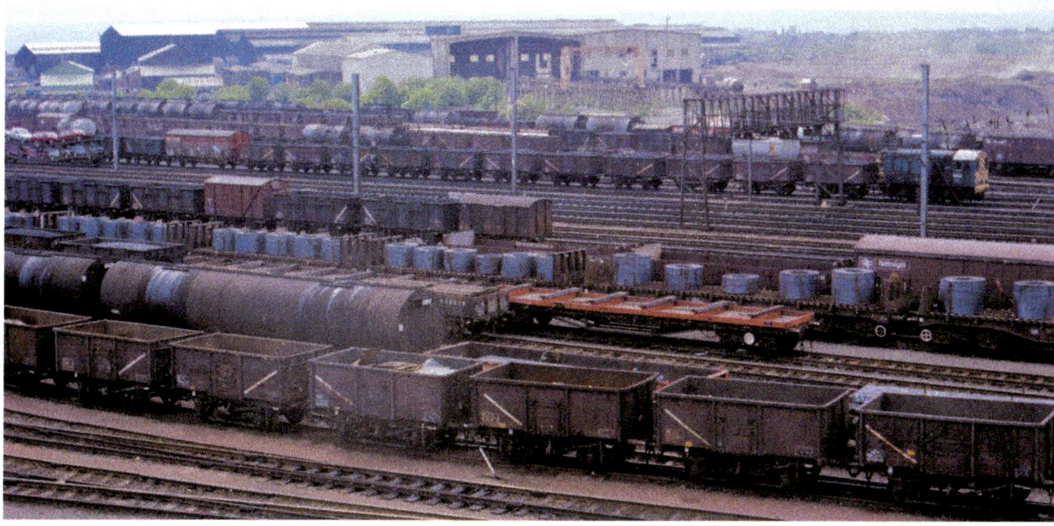

No. 08326 (ML) at Mossend Yard, near Bellshill, shunting wagons in preparation for a southbound departure later in the day. This yard is now a shadow of itself. Taken May 1982.

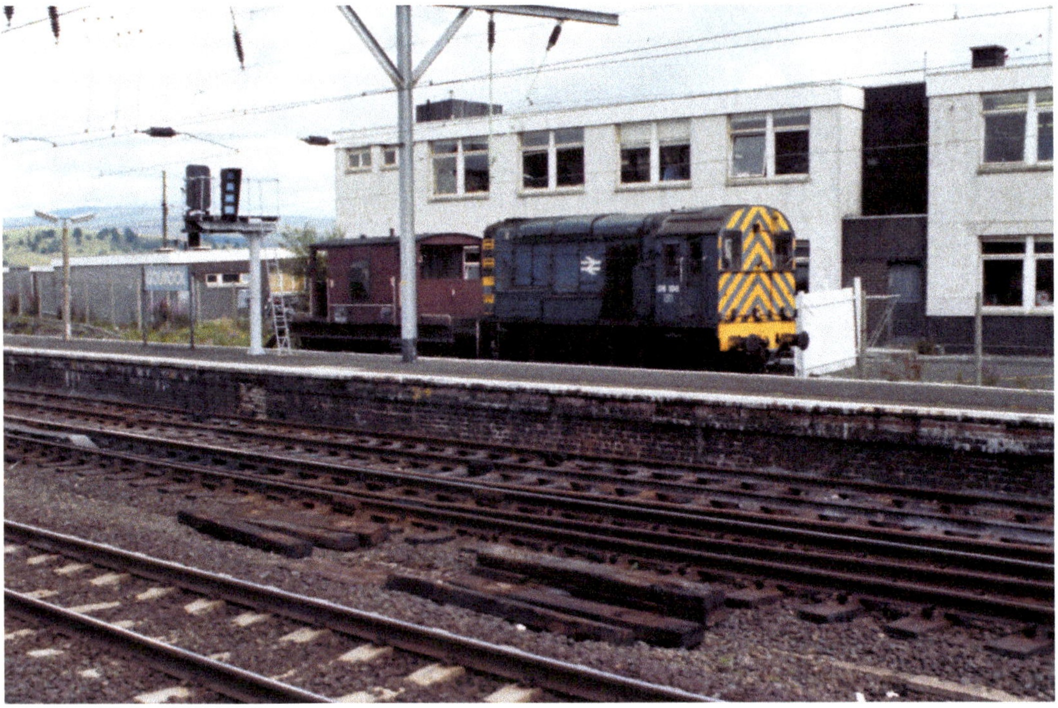

No. 08196 (ED) with a brake van stabled in the sidings at Gourock station. These sidings were removed in the mid-1980s. Taken August 1981.

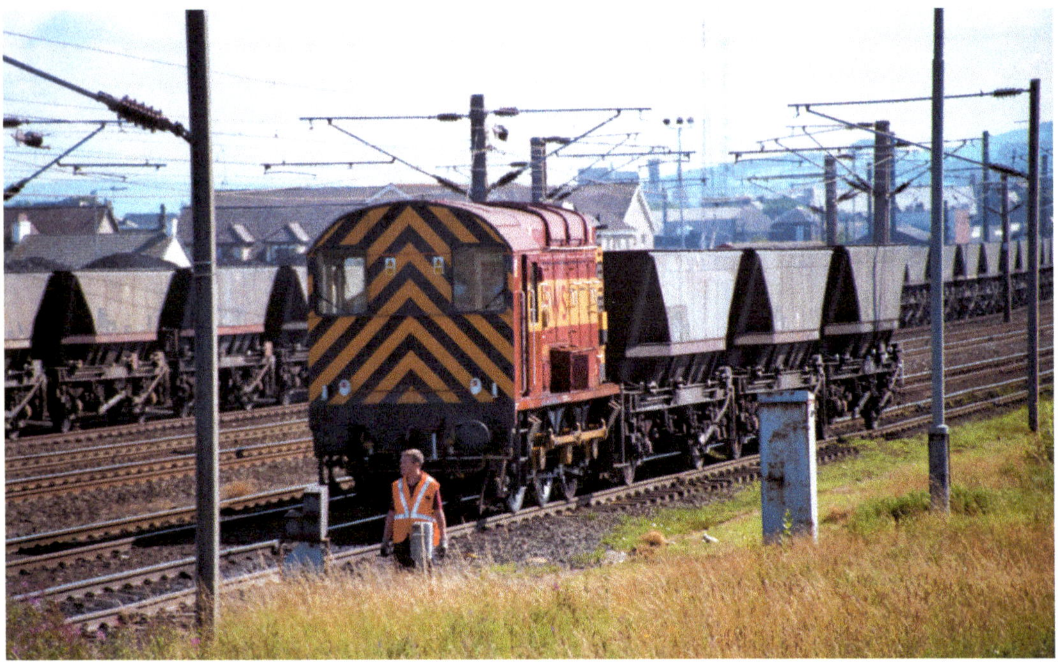

No. 08441 (ML) in EWS livery with three MGR wagons being shunted at Falkland Yard. Taken July 2001.